Studies in Physical Oceanography

Volume 2

Georg Wüst

Studies in Physical Oceanography

A Tribute to Georg Wüst
on his 80th Birthday

Volume 2

Edited by

ARNOLD L. GORDON

Lamont-Doherty Geological Observatory
of Columbia University
Palisades, New York 10964

GORDON AND BREACH SCIENCE PUBLISHERS
New York   London   Paris

Editor's Preface

The collection of articles composing this volume, represent the high esteem in which Georg Wüst is held by his colleagues. In each of the fields discussed in these studies Wüst has made important contributions; often his own articles are now considered the classical studies on which the present work is based. The contributions are ordered to match the seven major fields of study of Wüst, as given in the dedication by Gunter Dietrich.

In preparing this volume I unfortunately got off to a late start, so that the authors hadn't much time to prepare their contributions. This regrettably made it impossible for many of the "friends of Wüst" to contribute. There is also the distinct possibility I overlooked, due to ignorance, some oceanographers who would have liked to share in this tribute. This results from my contact with Wüst being limited to the last decade, while his activity in oceanography has spanned the last five and one half decades.

I attempted to limit the list of authors to those who know Wüst personally. Had I invited all oceanographers who are working in fields where Wüst has made important contributions to join in this "Tribute to Georg Wüst on His 80th Birthday" it would have filled many volumes and been far beyond my means to edit the collection

Georg Wüst celebrated his 80th birthday on 15 June 1970. At that time copies of the dedication and most of the papers were available and sent to Wüst at his home in Erlangen, Germany. Thus the main purpose of this volume, i.e. to present to Wüst on his birthday a collection of papers dedicated to him in friendship, was achieved nearly in total. However it will be near his 81st birthday that all the papers will be published for general distribution.

My contact with Wüst dates from 1961 when I began my graduate studies at Columbia University and Wüst was visiting professor. He remained in this capacity until 1964, during which time I worked closely with him on his Caribbean Sea studies. I learned from him that through combination of data and physical-mathematical analysis with imagination one can gain fairly detailed understanding of the ocean structure and circulation. The Wüst technique used with the computer and present day mathematical methods opens the door to a field which can be called modern descriptive oceanography. The results of this field form the basis on which the theoretical oceanographers must build.

Acknowledgements: In editing this volume I relied on the aid of Dr. Stephen Eittreim of Lamont-Doherty Geological Observatory, who looked after sending the contributions out for review during the summer of 1970 when I was at sea. His aid is much appreciated; it permitted progress to be made during the summer. Thanks are also due to the many oceanographers who reviewed the manuscripts. Their suggestions were most valuable in weeding out the minor inconsistencies or improper wording which so often creep into manuscripts and escape the view of the authors who are so close to their work. The secretarial work of Mrs. Jeanne Stolz (who also was Prof. Wüst's secretary during his days at Columbia University) is much appreciated. I also thank my wife, Susan, for her understanding during the times I spoke of nothing but the "Wüst Volume".

Special thanks go to the director of Lamont-Doherty Geological Observatory, Dr. Maurice Ewing, who was responsible for bringing Prof. Georg Wüst to Columbia, hence making it possible for direct contact of the scientists and students at Columbia with Georg Wüst.

ARNOLD L. GORDON

Palisades, New York

Table of Contents

Table of Contents ix

VOLUME 1

Spreading of Antarctic Bottom Waters II*

ARNOLD L. GORDON

Lamont-Doherty Geological Observatory of Columbia University
Palisades, N. Y. 10964

Abstract   The general spreading of bottom water by advective and diffusive transports is determined from distribution of bottom potential temperature, salinity and oxygen. The area south of 40° S between 130° W westward to 110° E is studied mainly with USNS Eltanin Cruises 27, 32–37, 39, 41, 42 and 43 data. The bottom water is composed of a mixture of three basic water types: lower Circumpolar Deep Water, water type B which is a cold relatively fresh ($34.66^o/_{oo}$) variety of Antarctic Bottom Water and water type C, a cold, but high salinity ($34.75^o/_{oo}$) variety of Antarctic Bottom Water. Type C is believed to be derived from the Ross Sea Shelf Water. It is found in the eastern and southern South Indian Basin, an area just northwest of the Ross Sea and in the southwestern corner of the Southeastern Pacific Basin. Water type B is more common. It is similar to the Bottom water observed leaving the Weddell Sea; however, other sources are suspected.

The bottom water spreading is basically clockwise in the basins between Antarctica and the mid-ocean ridge and from west to east in the basins to the north of the mid-ocean ridge. A flow from the South Indian Basin to the southern part of the Tasman Basin occurs near 145° to 155° E. This water eventually crosses the Macquarie Ridge to enter the Southwest Pacific Basin.

INTRODUCTION

Description of the properties of bottom water and their circulation has long been a subject of study. Distribution of the bottom temperature and concentrations of variously dissolved substances are used to detect the regions of the world ocean which act as source areas for the abyssal waters. These distributions define the major paths of spreading of bottom water throughout the world ocean. Knowledge of the quantity of water involved in the sinking process and in regions of compensating deep upwelling are used to determine the magnitude of ocean turnover and the extent of interaction of abyssal

---

* Lamont-Doherty Geological Observatory Contribution Number 1596.

waters with the surface water and atmosphere. Understanding the oceanography of the benthic layer of the ocean water aids in our interpretation of
the abyssal sedimentation record and the geological history of the earth.

The study of bottom water circulation would be best accomplished by the
direct recording of bottom currents over long intervals of time at numerous
positions. Simultaneous measurement of the physical and chemical properties of the water should also be made. Such a project would involve great
effort and cost. An alternative method is to measure easily determinable near
bottom parameters and deduce the bottom circulation pattern from their
distribution. This core method, developed by Wüst (1935a) has enabled
oceanographers to determine the general spreading (by advection and
diffusion) of bottom waters of the ocean. The most impressive of these
studies is reported in the numerous papers of Professor G. Wüst (1933, 1935b,
1937, 1938, 1957, 1964). Studies of spreading patterns by these methods are
consistent with the theoretical models developed by Stommel and Arons
(1960) and more recently by Kuo and Veronis (1970).

In recent years the greater amount of near-bottom hydrographic data has
allowed more detailed distributions of bottom parameters to be constructed
(Gordon, 1966, 1967; Reed, 1969; Gordon and Gerard, 1970; Reid and
Nowlin (in press), and others).

The subject of this paper is the description of bottom waters and their
circulation in a region south of 40° S from 130° W to 110° E (see Figure 1).
This area includes the major working area while in Antarctic Waters of the
U.S.N.S. *Eltanin* during the period, January, 1967 to June, 1970 (Cruises 27,
32–39, 41, 42 and 43 and a few earlier cruises, 16 and 24). The number II is
introduced into the title to indicate that this study is an extension of the early
work by the author (1966) which deals with the same subject (though different title) in the Southeastern Pacific (east of 170° W) and Scotia-Weddell Sea
region.

The hydrographic data gathered by the *Eltanin* is available from Jacobs
and Amos, 1967 for Cruises 24, 26 and 27; from Jacobs, Bruchhausen and
Bauer (1970) for Cruises 32–39 and from J. L. Reid of Scripps Oceanographic
Institution for Cruise 41. Cruise 42 and 43 data is still in preliminary form
and will appear in a data report. Data from *Eltanin* Cruise 16 (Jacobs, 1966)
were also used.

THE DATA

The bottom potential temperature, t_p, salinity and oxygen are used in this
study. Adiabatic correction is determined from the tables of Helland-
Hansen (Wüst, 1961). *Eltanin* data are divided into four groups according

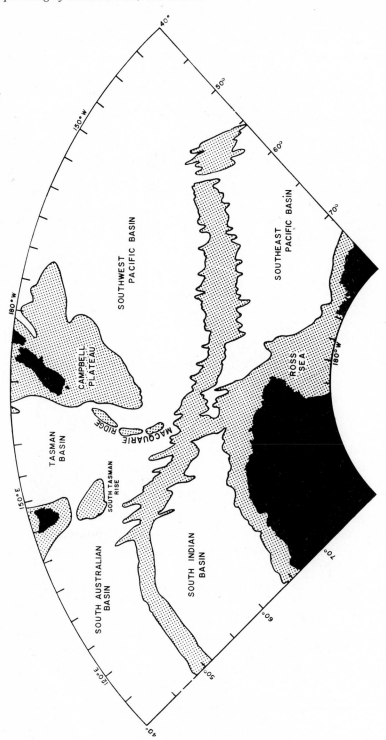

Figure 1 The area included in this study

to the deepest observation in relation to the sea floor: data from 0 to 100 m from the sea floor (sonic depth), 101 to 200 m, 201 to 300 m and 301 to 400 m. The regional near bottom vertical gradients were inspected to determine the usefulness of data which is greater than 200 meters from the sea floor. The data from 0 to 100 meters was given more weight in the contouring process. Table 1 lists the number of stations in each depth category.

Table 1

Depth	*Eltanin* Cruises													
Interval (m)	16	26	27	32	33	34	35	36	37	39	41	42	43	Total
0–100	4	2	19	2	2	12	8	19	31	6	13	5	5	128
101–200	0	0	3	0	0	6	4	13	2	2	3	0	0	24
201–300	1	0	0	0	2	1	7	3	0	0	1	0	0	15
301–400	0	0	0	0	1	0	1	0	1	1	1	0	0	5

Noted:
  [1]) Cruises 27, 32 have much data in Ross Sea, which is not treated in this study (see Jacobs, Amos, Bruchhausen, in press).
  [2]) Most of Cruise 33 data is east of 130° W.
  [3]) Ship's data other than *Eltanin* totals in each category, respectively: 21, 15, 16, 23; therefore, a total of 247 stations are used in this study.

In addition to *Eltanin* data, all available bottom data were obtained from the National Oceanographic Data Center (NODC). There are in all seventy-five stations obtained by ships other than *Eltanin*. The *Discovery* data comprises 46% of that total.

This study overlaps the 1966 work (from 130° W to 170° W) to yield a more useful overall view of the southwest Pacific area and the influence of the Ross Sea on the deep ocean. Bottom t_p and O_2 contours between 130° W to 170° W have been modified from the earlier work with the aid of *Eltanin* data from Cruises 24 and 33, 42 and 43. The bottom salinity distribution in the overlap area is constructed from the original data since bottom salinity was not included in the 1966 work as there are only minor bottom salinity gradients in the southeastern Pacific Ocean. In the region presently under consideration, the salinity pattern is more meaningful in deducing bottom circulation because of its greater range and relatively strong bottom gradients.

TEMPERATURE–SALINITY RELATIONS

The relationship of bottom potential temperature to salinity is shown in Figure 2. This diagram is constructed from data with depths within 100 m of the sonic depth. The shape of the $t_p - S$ curve suggests that the bottom water is to varying degrees a blend of three basic water types, labeled A, B, and C on Figure 2. These water types have the properties shown in Table II. The bottom water falls into two categories: a blend of either types A and B or B and C. There appears to be no water blending types A and C.

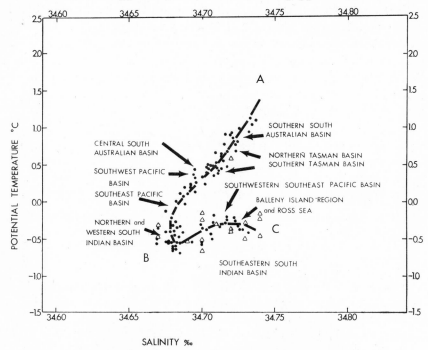

SALINITY ‰

Figure 2 Relationship of bottom potential temperature to bottom salinity for all data points within 100 m of sonic depth. The geographic distribution of the t_p/S points are shown

Type A is the lower Circumpolar Deep Water (CDW) (Gordon, 1967). Type B is very similar to the Antarctic Bottom Water observed leaving the Weddell Sea (Mosby, 1934, p. 79 and Gordon, 1967, Plate 8). It is low in salinity and temperature, and high in oxygen (near 6 ml/l). This bottom water, generally considered to be the sole type of Antarctic Bottom Water (AABW), is widespread (Wüst, 1933). Water type C is also cold, and high

Table 2

Water Type	t_p °C	$S^0/_{00}$
A	+1.7	34.74
B	−0.8	34.66
C	−0.5	34.75

in oxygen (5.5 to 6.0 ml/l), but unlike type B, its salinity is high. It is likely that this water flows from the salty Ross Sea Shelf water (Gordon, 1970; Jacobs, Amos and Bruchhausen, in press). Gordon (1970) suggested, on the basis of very low temperatures of the Shelf water that a large part of its volume results from freezing to the underside of the thick Ross Ice Shelf. Therefore, it is probable that the high salinity water type C is an Antarctic Bottom Water of a different origin than type B.

Types B and C are two distinctly different varieties of Antarctic Bottom Water. Their different nature suggests different methods of production. It is important in further discussions of Antarctic Bottom Water to indicate the type which would indicate its mode of production. Gordon (1970) suggested a nomenclature for AABW. It is too early to assign such names to water types B and C, but eventually we hope to be able to do so.

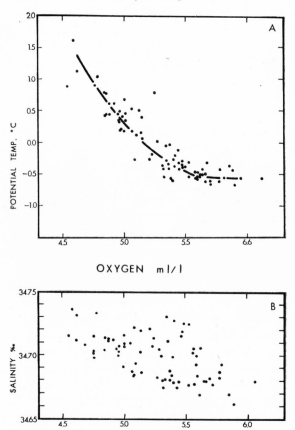

Figure 3   Relationship of bottom potential temperature to bottom oxygen (a) and bottom salinity to bottom oxygen (b) of *Eltanin* data points within 100 m of sonic depth

OXYGEN

The relation of bottom potential temperature and salinity to oxygen is shown in Figures 3a and 3b, respectively. The low temperature is related to high oxygen. The relation of salinity to oxygen is a bit more complicated. The highest oxygen is associated with salinities of less than 34.69 $^o/_{oo}$ (type B), but relatively high oxygen values are associated with the high salinity water type C. This indicates that the water comprising these bottom water types has been recently near the sea surface or perhaps, as suggested by Jacobs *et al.*, (in press), picked up oxygen in a melting process of the ice shelf.

REGIONAL DISTRIBUTION OF t_p, S, O$_2$

The distributions of bottom potential temperature, salinity and oxygen are given in Figures 4, 5 and 6, respectively. Only *Eltanin* oxygen data is used since there are too few *Discovery* bottom data to derive a reliable correction factor as in the case of the Southeastern Pacific Basin (Gordon, 1966). The positions on a t_p/S diagram of the near-bottom data from various geographic locations are shown in Figure 2.

The warmest water occurs along the northern flank of the mid-ocean ridge south of Australia. Here the core layer of the lower CDW approaches the sea floor. The water within the deeper segments of the South Australia and South Indian Basins is colder and falls on a t_p/S line connecting lower CDW and water type B. The southeastern part of the South Indian Basin clearly displays the high salinity type C.

A connection of the $-0.4°C$ isotherm across the Balleny Plateau cannot be made from available data, but the large near-bottom temperature gradients suggest that a thin cold layer may exist. A cold water feature extends from the Ross Sea into the Southeastern Pacific Basin (Bellingshausen Basin) and has been cited by Gordon (1966) as evidence for a bottom water contribution from the Ross Sea. From its position on the t_p/S diagram, it is believed to be of type C, but in lesser concentration than found in the South Indian Basin. Since bottom water of type B also occurs in the Southwestern Pacific Basin, it is possible that either the Ross Sea is a producer of this type of bottom water, or it is derived from the Weddell Sea via the southern Drake Passage (Gordon, 1967). Figure 7 shows the typical vertical distribution of temperature and salinity in a region with AABW of type C.

The cold water (as low as $-0.6°C$) found in the southern South Indian Basin is basically water type B which has been diluted with lesser amounts of warmer water than has the bottom water within the South Australian Basin (Figure 2). Is it possible that the former came from the Weddell Sea

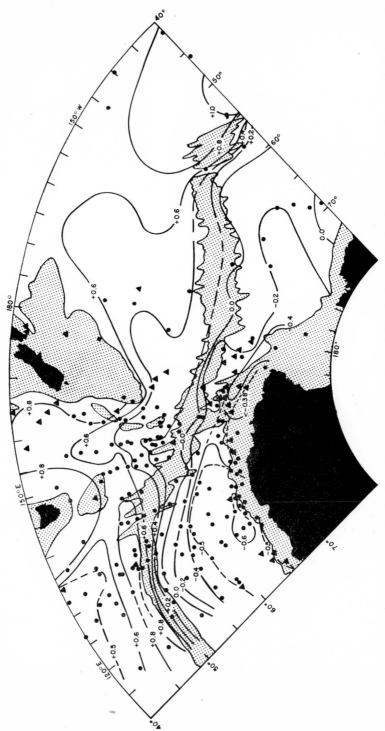

Figure 4 Bottom potential temperature distribution. The data are positions of *Eltanin* data and triangles are positions of other ships' data used in this study

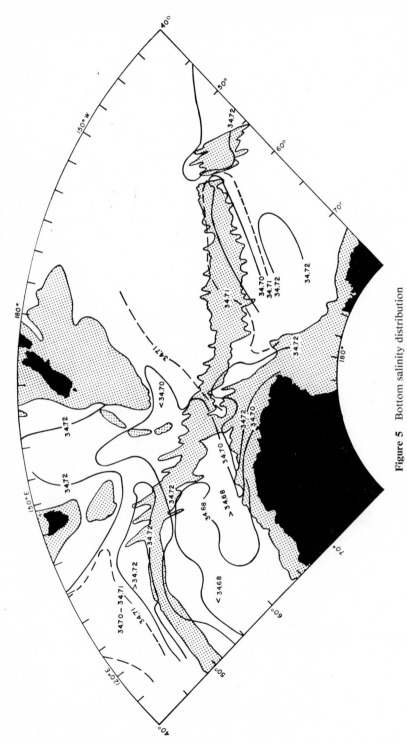

Figure 5 Bottom salinity distribution

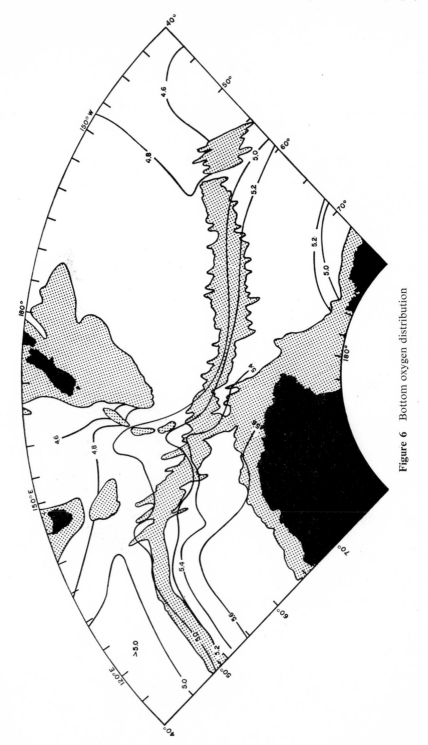

Figure 6 Bottom oxygen distribution

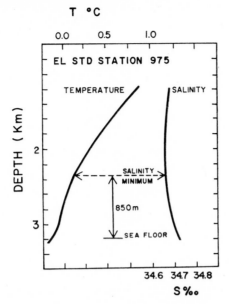

Figure 7   Corrected and smoothed analog trace of temperature (in situ) and salinity
versus depth at 65°16′ S and 156°03′ E (*Eltanin* Cr.37)

in such a concentrated form? If not, a third source region of type B must
exist (besides the Weddell and Ross Seas). *Eltanin* coverage west of 120° E
in the Indian Ocean sector of Antartic waters, or perhaps complete analysis
of the great quantities of data collected along the Adelie Coast (Cruise 37)
may answer this question.

BOTTOM CURRENT PATTERNS

The distribution of bottom properties depends on three things: 1) advection;
2) eddy diffusion (both lateral and vertical); and 3) a time dependent term,
such as oxygen consumption, radioactive decay or the heating of bottom
waters by geothermal heat flux. For the commonly measured bottom properties, the third term can be neglected in areas of limited extent since the
advection and diffusion transports far outweigh its effects in governing the
distribution of these parameters. For short-lived isotopes, the non-conservative term is important. If the spread of bottom properties were accomplished only by advection, streamlines would be parallel to the isolines
(water properties are constant along flow lines). If the spreading were
governed only by diffusive transports, the properties would spread at right
angles to their isolines. Since the spread of bottom waters is accomplished
mainly by a combination of advection and diffusion, deviation between the

streamlines and isolines is expected. The classical advection-diffusion equation for steady state conditions of a conservative property, S (concentration) and coefficient of eddy diffusion independent of space coordinates is given by

$$u\frac{\partial S}{\partial x} + v\frac{\partial S}{\partial y} + w\frac{\partial S}{\partial z} = \frac{A_x}{\varrho}\frac{\partial^2 S}{\partial x^2} + \frac{A_y}{\varrho}\frac{\partial^2 S}{\partial y^2} + \frac{A_z}{\varrho}\frac{\partial^2 S}{\partial z^2}$$

Arranging the coordinate system so that the x-axis is directed along the velocity u (making $v = x = 0$) and assuming the diffusive transport in the x-direction is much smaller than the advective transport in that direction leads to

$$\frac{Ay}{\varrho u}\frac{\partial^2 S}{\partial y^2} + \frac{Az}{\varrho u}\frac{\partial^2 S}{\partial z^2} = \frac{\partial S}{\partial x}$$

The values of abyssal mixing coefficients are approximately 10^7 cm²/sec for Ay/ϱ (Kuo and Veronis, 1970) and 10 for Az/ϱ (Broecker, Cromwell and Li, 1968) based on near-bottom radon measurements. Mean bottom velocities are usually small being less than 10 cm/sec with perhaps an average value near 1 cm/sec. From Figure 4, the lateral second derivatives for the salinity field can be determined and from the hydrographic stations the vertical derivatives can be found. In order to calculate the lateral second derivative, the angle between streamline and isohalines must be known. Since it is this that we are attempting to find, I will use the maximum second derivative values (i.e. parallel to the salinity gradient) which would accentuate the diffusive transport. This results in a larger $\partial S/\partial x$ value for each set of Ay and u or larger angular difference of stream-lines to isohalines. The largest lateral second derivative of salinity distribution is of the order 10^{-19} cm^{-2} along the flanks of the mid-ocean ridge and 10^{-20} cm^{-2} in the central areas of the basin. These values are given to only an order of magnitude since the construction of the isohalines was a partially subjective procedure. The vertical second derivative is 10^{-14} to 10^{-15} cm^{-2}.

The value of the change of salinity along the x-axis, $\partial S/\partial x$ can be approximated using the numerical values described above. For $u = 10$ cm/sec a value of 10^{-13} cm^{-1} is found (the lateral diffusion term contributes most of the diffusive transport). A gradient of 10^{-13} cm^{-1} requires a distance of 2000 km of water flowing along a streamline to change its salinity by diffusion $0.02°/_{00}$ (the contour interval used in this study). For the case of a bottom velocity of 1 cm/sec, the required distance for an alteration of $0.02°/_{00}$ is only 200 km.

The average magnitude of the instantaneous bottom current in the area of study is near 5 cm/sec (Eittreim *et al.*, this volume). Therefore, a distance of about 1000 km of travel is needed for the flow to cross from one isohaline to the next.

The angle between streamline and isohaline depends on the ratio of the salinity gradient perpendicular and parallel to the streamline, $\partial S/\partial x$ and $\partial S/\partial y$. The relation is:

$$\sin \beta = \frac{\Delta \zeta}{\Delta x}$$

when β is the angle between the streamlines and isohalines, $\Delta \zeta$ and Δx, are the distance over which the salinity changes by $0.02\,^0/_{00}$ parallel to the salinity gradient and parallel to the flow, respectively. The value of Δx is calculated from the advection-diffusion equation and $\Delta \zeta$ is measured. In general, $\Delta \zeta$ is of the order of 100 to 200 km near the ridges and of an order near Δx over the basins where gradients are small. If Δx is 1000 km, the angle β is about 5° near ridges and near 90° in the central regions.

The current pattern deduced from the distributions shown in Figures 4, 5, and 6 is displayed in Figure 8. In view of the above discussion, the current pattern is quantitatively constructed so that the flow is mostly parallel to the isolines near the ridges and continental margins and at a larger angle in the central parts of the basins. In general, the flows in the two southernmost basins (South Indian Basin, Southeast Pacific Basin) are clockwise while the flow in the two northern basins is generally from west to east.

There is northward penetration of the cold bottom waters from the South Indian Basin across the mid-ocean ridge into the Tasman Basin at approximately 145° to 155° E longitude. This region is seismically active (Sykes, 1966) and may be cut by deep fractures. The bottom waters in the Tasman Basin south of 45° S latitude are very similar to those of the South Indian Basin and dissimilar, in that they are colder and fresher than the waters of the South Australian Basin. The waters of the South Australian Basin appear to spread eastward through the passage to the south of the South Tasman Rise and then turn sharply to the north on entering the Tasman Basin. The water in the eastern sections of the basin may represent the southward return of this water and most likely flows into the Southwest Pacific Basin via the northernmost deep passage in the Macquarie Ridge. Since these waters are not clearly seen in the Southwest Pacific Basin, it is possible that they are confined to a relatively narrow and swift flow compressed against the sides of the Campbell Plateau. The presence of ripple marks on the sea floor in this vicinity (Camera Station #42, Cruise 36) suggests strong bottom currents.

Some of the cold water of the Hjort Trench flows through a passage-way in the Macquarie Ridge at approximately 56° S latitude and enter the South-

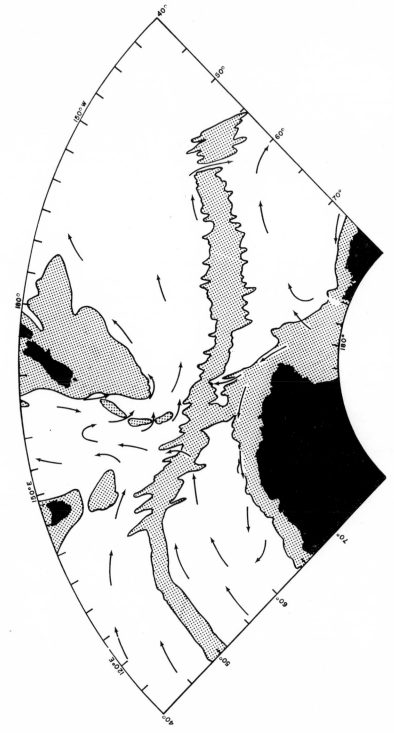

Figure 8 Proposed pattern of circulation of bottom waters deduced from parameter distribution shown in Figures 3, 4, and 5

west Pacific Basin. The thermometric depth of this passage is approximately 4000 m (Cruise 37 Final Cruise Report, Gordon, 1969). The combined flow of water flowing through the passage of the Macquarie Ridge and around the southern end of the Ridge enters the Southwest Pacific Basin* and contributes to the western boundary undercurrent in the South Pacific (Warren, 1970). It is probable that not all of this water enters the western boundary undercurrent but some flows across the Southwestern Pacific Basin and enters the Southeast Pacific Basin by way of the USARP Fracture Zone (Gordon, 1967).

Since the distribution of bottom parameters responds to the advective-diffusive processes, they represent an average bottom circulation. High frequency variations would have little effect on the distribution of bottom parameters. For example, if a moderate current pattern (velocities of 5–10 cm/sec) reversed itself, it would take weeks before the distribution of bottom parameters would be altered enough to be observed (this value depends upon the horizontal gradients of bottom parameters and the accuracy of the measurements). Therefore, the current pattern deduced from the distributions of temperature, salinity, oxygen or other parameters represents an average condition existing over a period of weeks, or months and gives no indication of the high frequency periodic or transient motions. It is advisable to keep this in mind when comparing this current pattern with instantaneous current measurements (Eittreim *et al.*, this volume) or perhaps even with twentyfour hour time-series current measurements.

CONCLUSIONS

The distribution of bottom potential temperature, salinity and non-conservative properties as oxygen is used to deduce the water mass composition of the bottom waters, and yield a qualitative picture of the bottom current pattern. The information gained from this data describes only the mean or, in a sense, climatic bottom conditions. High frequency variations (less than weekly or perhaps monthly fluctuations) go unnoticed in the large scale structure of the bottom waters.

The distribution of bottom properties in the study area indicates that Antarctic Bottom Water is not a simple dilution of one particular water type of bottom water and Circumpolar Deep Water, but that at least two different types of Antarctic Bottom Water are represented. Both are cold and high in

---

* The Macquarie Ridge region was investigated during *Eltanin* Cruise 44, (June–August, 1970). These data form the basis for a more detailed description of the interaction of the ridge with the Antartic Circumpolar Current.

oxygen content, but have different salinities. The low salinity appears to be more widespread than the high salinity variety.

The low salinity type B of Antarctic Bottom Water is similar to that variety observed leaving the Weddell Sea. The relatively concentrated form of type B found in the South Indian Basin and Southeast Pacific Basin suggests a source of this water type in closer proximity to each of these basins than the Weddell Sea.

The high salinity variety (type C) of Antarctic Bottom Water is derived from Ross Sea Shelf Water (found in the southwestern Ross Sea). Its fairly rapid dilution in the southeastern section of the South Indian Basin and Southeast Pacific Basin indicates either very large coefficients of eddy diffusion or more probably a small amount of type C produced.

The current pattern in the region is basically west to east flow, with a westerly flow along the Antarctic Continental margins. A northward flow across the mid-ocean ridge near 145°–150° E introduces cold bottom water into the Tasman Basin. This water eventually crosses or outflanks the Macquarie Ridge to the south and enters the Southwestern Pacific Basin and contributes to the South Pacific western boundary undercurrent.

Acknowledgments

The comments of S. S. Jacobs and S. Eittreim are appreciated. The *Eltanin* data collection and analysis program is funded by the Office of Polar Programs of the National Science Foundation. The U.S. Atomic Energy Commission (AT 30-1 (2663) contract) supported part of the field work of numerous *Eltanin* cruises.

References

Broecker, W. S., J. Cromwell and Y. H. Li, Rates of vertical eddy diffusion near the ocean floor based on measurements of the distribution of excess 222Rn. Earth and Planetary Science Letters 5, 101–105 (1968).

Eittreim, Stephen, A. L. Gordon, M. Ewing, E. M. Thorndike and P. Bruchhausen, The Nepheloid Layer and Observed Bottom Currents in the Indian-Pacific Antarctic Sea. In: *Studies in Physical Oceanography—A Tribute to Georg Wüst on His 80th Birthday*. (1971).

Gordon, A. L., Potential temperature, oxygen and circulation of bottom water in the Southern Ocean. *Deep-Sea Res.* **13**, 1125–1138 (1966).

Gordon, A. L., Structure of Antarctic waters between 20° W and 170° W. Antarctic Map Folio Series No. 6 (V. Bushnell, Editor) Amer. Geograph. Soc. (1967).

Gordon, A. L., Recent physical oceanographic studies of Antarctic waters. In: *Antarctic Research* (Ed. by L. Quam), Amer. Assn. Adv. Sci., Washington, D. C. (1970).

Gordon, A. L. and R. D. Gerard, North Pacific Bottom potential temperature. In: *Geological Investigation of the North Pacific*, GSA Memoir 126 (1970).

Jacobs, S.S., Physical and chemical oceanographic observations in the Southern Oceans, USNS *Eltanin*, Cruises 16–21, 965, L-DGO Tech. Rept. 1-CU-1-66 (1966).

Jacobs, S.S. and A. Amos, Physical and chemical oceanographic observations in the Southern Oceans, USNS *Eltanin* Cruises 22–27, 1966–67, L-DGO Tech. Rept. 1-CU-1-67 (1967).

Jacobs, S.S., A. Amos and P. Bruchhausen, Ross Sea oceanography and Antarctic bottom water formation. *Deep-Sea Research* 17 (6): 935–962 (1970).

Jacobs, S.S., P.M. Bruchhausen, and E.B. Bauer, *Eltanin* Reports Cruises 32–36, hydrographic stations, bottom photographs and current measurements, Lamont-Doherty Geol. Obs., Palisades, N.Y. (1970).

Kuo, H. and G. Veronis, Distribution of tracers in the deep oceans of the world. *Deep-Sea Research* 17 (1), 29–46, (1970).

Mosby, H., The waters of the Atlantic Antarctic Ocean. Sci. Res. Norweg. Antarc. Exped., 1927–28, 11, 1–131, (1934).

Reed, R.K., Deep water properties and flow in the Central North Pacific. *J. Mar. Res.* 27 (1), 24–31 (1969).

Reid, J.L. and W.D. Nowlin, Transport of water through the Drake Passage. *Deep-Sea Research* (1970, in press).

Stommel, H. and A.B. Arons, On the abyssal circulation of the world ocean—II. An idealized model of circulation pattern and amplitude in oceanic basins, *Deep-Sea Research* 6, 217–233 (1960).

Sykes, L., Seismicity of the South Pacific Ocean. *J. Geophys. Res.* 68 (21), 5999–6006 (1963).

Warren, B., Antarctic contribution to water circulation. In: *Antarctic Research* (Ed. by L. Quam), Amer. Assn. Adv. Sci., Washington, D.C. (1970).

Wüst, G., Das Bodenwasser und die Gliederung der atlantischen Tiefsee, (Wiss. Erg. d. D.A.E. *Meteor* 1925–27, Bd. VI (1) Lfg.) Berlin, 1–107 (1933).

Wüst, G., Die Stratosphäre des Atlantischen Ozeans (Wiss. Erg. d. D.A.E. *Meteor* 1925–27, Bd. VI (2) Lfg. Berlin, 1–288 Mit Atlas) (1935a).

Wüst, G., Die Ausbreitung des antarktischen Bodenwassers im Atlantischen und Indischen Ozean (Zeits. f. Geophysik, 11, H.1/2 Braunschweig, 1–49) (1935b).

Wüst, G., Bodentemperatur und Bodenstrom in der pazifischen Tiefsee (Veröff. Inst. f. Meereskunde, N.F. Reihe A, H.35, Berlin, 1–56) (1937).

Wüst, G., Bodentemperatur und Bodenstrom in der atlantischen, indischen und pazifischen Tiefsee (Gerlands Beiträge zur Geophysik, Bd. 54 (1), 1–8) (1938).

Wüst, G., Stromgeschwindigkeiten und Strommengen in den Tiefen des Atlantischen Ozeans unter besonderer Berücksichtigung des Tiefen- und Bodenwassers (Wiss. Erg., Deutsch. Atl. Exped. *Meteor* 1925–27, Bd. VI (2), Teil 6, Lfg. 1–160, 6 Abb., 36 Taf. u. 1 Beil., Berlin) (1957).

Wüst, G., Tables for rapid computation of potential temperature. Tech. Rept. CU-9-61 AT (30-1) 1808, Lamont Geol. Obs. (1961).

Wüst, G., Stratification and circulation in the Antillean-Caribbean basins. VEMA Research Series No. 2, Columbia University Press, N.Y., 1–201 (1964).

The Nepheloid Layer
and Observed Bottom Currents
in the Indian–Pacific Antarctic Sea*

STEPHEN EITTREIM, ARNOLD L. GORDON,
MAURICE EWING, EDWARD M. THORNDIKE
AND PETER BRUCHHAUSEN

Lamont-Doherty Geological Observatory, Columbia University
Palisades, New York 10964

Abstract A nepheloid layer of roughly 1 km thickness is associated with the bottom waters of the basins south of the mid-ocean ridge in the Indian-Pacific Antarctic region between 120° E and 120° W. A downward increase in light scattering intensity of about a factor of six occurs in this layer. The mid-ocean ridge is breached by the turbid, cold bottom water only at its intersection with the Macquarie Ridge. The water which produces the highest light scattering intensities is Antarctic Bottom Water (AABW) with high oxygen content and it occurs close to the Antarctic continent in the South Indian Basin. Apparently the nepheloid layer is fed by particles suspended in the dense shelf water as it sinks to form new AABW. Turbulence in the AABW then maintains the particles in suspension as the water mass circulates around and through the deep basins. This turbulence accompanies the irregular bottom currents which often reach velocities of over 15 cm/sec. The short duration current measurements reported here show a dominance of irregular motion, with a mean flow clockwise in the South Indian Basin. The depression of the top of the nepheloid layer in the center of this basin is a reflection of this clockwise flow of nepheloid AABW.

INTRODUCTION

During *Eltanin* cruises 32–35, 37 and 39, a Lamont–Doherty tripod system (Thorndike and Ewing, 1969) was used to investigate the water column and the environment of deep-sea sedimentation. A nephelometer to measure a vertical profile of turbidity, a bottom camera and a bottom current meter

---

* Lamont-Doherty Geological Observatory Contribution Number 1601.

are combined in the tripod system. Over 80 such lowerings were made in the region from 120° W westward to 120° E. In this paper, the nephelometer and bottom current meter observations are discussed and related to the preceding hydrographic study of this region, by Gordon, a study hereafter referred to as "II".

The mid-ocean ridge, Tasman and Macquarie ridges divide this region into the following five basins, proceeding clockwise, from south of Australia (Figure 1): South Australian, Tasman, Southwest Pacific, Southeast Pacific (or Bellingshausen) and South Indian.

The bottom circulation is strongly influenced by sources of cold, dense water from the Antarctic shelf. Besides the Weddell Sea as one of these source areas, it has been demonstrated (II; Gordon, 1966; Jacobs, *et al.*, in press) that the Ross Sea may be another source of bottom water which spreads to the Bellingshausen and South Indian basins. As these bottom waters spread away from Antarctica, westward flowing currents along the continental margin of Antarctica are expected due to Coriolis force deflection.

TRIPOD SYSTEM

The tripod is lowered to the bottom with a hydrographic winch and is placed on the bottom for a total of about one hour, during which time several "lift-offs" are made to check for consistency in the current measurement between the different periods of measurement on bottom (called "hits") and also to obtain bottom photographs in several different locations. With a ship drift of 1 knot and "lift-off" time duration of five minutes (a maximum value), the distance between hits would be 150 m.

The current meter is suspended from the apex of the tripod and the sensing elements are at a distance of approximately 0.75 m above the sea floor. Current magnitude is measured by the deflections of three thin aluminum pendulums. These deflections and the positions of a direction vane and compass are photographically recorded. A complete discussion of the pendulum current meter including its calibration and data reduction methods is given in Thorndike and Ewing (1969). The methods which they outlined are designed to minimize uncertainties and were followed in the reduction of these data. Bottom photographs are taken simultaneously with the current meter photographs and the former are used to check the stability of the tripod on bottom.

The lower limit of sensitivity of the meter is about 1 cm/sec and velocities below 2 cm/sec are somewhat unreliable. The maximum velocity which can be recorded, which is the point at which the pendula approach a horizontal attitude, is about 15 cm/sec. There are two compasses on the tripod, one in the field of view of the current meter camera and a second in the field of view of the bottom camera. Both compasses are used and the average difference between the two is reported. It is especially critical to have such a check on the compass reading in this area which is close to the south magnetic pole where the horizontal component of the magnetic field is very weak. At six stations there was substantial disagreement between the two compasses (greater than 60°) and in these cases the bottom camera compass reading was used since this is considered to be the more reliable (Thorndike and Ewing, 1969). Correction for magnetic declination was made using U.S. Hydrographic Office chart no. H0 1706-5.

The nephelometer is a photographically recording device which records in situ small angle (approximately 15°) light scattering continuously while the tripod is lowered to the bottom and raised again. The distance between light source and camera is 55 cm and the long axis of the nephelometer lies horizontally between support members of the tripod. A bourdon-tube pressure recorder records depths internally. Direct light from the light source is attenuated and recorded to monitor the light output. A detailed description of the nephelometer is given in Thorndike and Ewing (1967).

CURRENT METER RESULTS

Time series measurements of bottom currents made over extended periods (on the order of a day or longer) have demonstrated that in the deep sea irregular motions are a common, if not the dominant, type of water movement. Reid and Nowlin (in press), for example, have found that in the northern Drake Passage the amplitude of diurnal variations in near-bottom currents is greater than the amplitude of the mean eastward flow. The steady state flow which is superimposed on these irregular motions is the main point of interest and is often difficult to ascertain from spot measurements. However, if a large enough number of spot measurements are made in a region, at random times, a mean flow pattern, if such exists, should make itself evident.

The measurements reported here are essentially spot measurements, being made over a time interval of about one hour or less, and hence are subject to the above limitation. Figure 1 shows the mean magnitude and direction of the measured current, averaged over all "hits". The spread in measured magnitudes can be seen in Table 1 in which the magnitude for each hit is listed.

The 83 current measurements show the extreme irregularity of flow in these deep waters, even where stations are closely spaced, but they also reveal

Table 1 Current Meter Results

Eltanin 33

Station	Latitude S	Longitude W	Corrected Depth (m)	Measured Velocities on Successive Hits* (cm/sec)	Mean Velocity (cm/sec)	Direction	Difference between Two Compasses**	Duration (minutes)
1	60°34'	171°46'	4360	8	7.8	190	-29	20
2	66°59'	164°51'	3744	6, 8	7.0	100	-9	38
3	68°01'	159°52'	3726	9, 8, 6	7.6	290	8	25
4	68°49'	156°11'	4220	3, 3, 4	2.7	170	59	25
5	65°04'	139°48'	4432	3, 3, 4	3.2	160	30	32
6	67°03'	136°51'	4510	3, 2, 3	2.5	015	34	34
7	68°46'	134°05'	4360	4, 6, 7	6.6	195	-11	22
8	69°27'	130°37'	3136	0	0	—	—	35
9	69°33'	124°49'	2907	4, 6, 4	4.4	005	43	27
10	70°05'	122°17'	3704	2, 0, 0	1.8	095	-9	15
11	70°06'	119°45'	2891	3, 3	3.2	230	-19	10
12	68°53'	120°14'	4092	5, 6, 7, 4	5.5	300	38	30
13	67°59'	119°59'	4310	> 15, 13, 12	~14	295	-4	60
14	65°06'	120°09'	4948	9, 11, 8, 9	8.8	270	37	35
15	63°11'	120°04'	5033	7, 8, 8, 9	7.9	160	-11	51
16	62°03'	119°51'	4159	7, 11	8.6	205	-23	10
17	61°02'	119°51'	5042	0, 0, 3	0	—	—	20
18	59°53'	119°38'	4547	3, 3, 3	3.2	150	-26	69
19	59°01'	119°54'	4776	0, 0	0	—	—	10
20	58°13'	120°02'	4292	9, 9, 6	7.6	255	-9	31
21	56°27'	119°47'	4410	10, 6, 7, 6, 8	7.2	310	-18	30
22	54°56'	120°01'	2760	10, 7, 7, 7	7.4	280	-3	50

* Rounded to nearest whole number.
** Angular difference: camera compass minus current meter compass.

Table 1 (cont.)

Eltanin 34

Station	Latitude S	Longitude E	Corrected Depth (m)	Measured Velocities on Successive Hits (cm/sec)	Mean Velocity (cm/sec)	Direction	Difference between Two Compasses	Duration (minutes)
1	55°53'	170°12'	5240	>15, >15, >15	>15	105	+22	35
2	58°01'	170°06'	5242	11, >15, >15, >15, >15, >15	>15	025	−35	58
4	60°08'	167°40'	4634	6, 5, 5, 6	5.1	105	−46	47
5	60°14'	159°53'	3740	>15, >15, >15, >15, >15	15	315	−49	50
6	57°25'	160°00'	3896	10, 7, 9, 8, 8	8.3	210	−3	52
7	56°08'	159°59'	3858	12, 10, 12	11.3	310	−8	42
8	51°23'	159°59'	3887	4, 4	3.6	030	−37	35
9	49°42'	160°07'	4448	6, 7, 7, 6	6.4	295	+1	54
10	38°15'	160°02'	4521	2, 1, 1	1.6	110	−18	28
12	48°05'	145°04'	3907	7, >15, 9	9.4	150	−31	29
13	51°08'	144°56'	3658	>15, 14, >15, >15	>15	200	+13	34
14	53°29'	144°59'	3015	3, 4, 5, 7	4.6	285	+53	49
15	58°08'	144°55'	3745	3, 3	3.1	185	−2	17
16	60°12'	144°47'	3906	12, >15	>15	100	−15	55
17	60°04'	140°04'	4447	0, 0, 0, 0	0	—	—	46
18	60°01'	134°51'	4566	8, 8, 7, 7, 8	7.7	020	+143*	58
19	57°56'	135°02'	4625	4, 4, 6, 6	5.1	115	−4	59
20	56°38'	135°11'	4054	2, 2	2.0	045	−152*	60
21	54°03'	135°09'	3954	10, 8, 7	8.5	160	+114*	77
22	52°02'	135°09'	3411	5, 4, 5, 6, 5	4.7	085	+46	61
23	50°21'	135°01'	3060	7	6.7	200	+27	10
24	45°00'	134°57'	4518	2, 0	0.8	170	+27	63

* Large difference between two compasses; camera compass used.

Table 1 (cont.)

Eltanin 35

Station	Latitude S	Longitude E	Corrected Depth (m)	Measured Velocities on Successive Hits (cm/sec)	Mean Velocity (cm/sec)	Direction	Difference between Two Compasses	Duration (minutes)
1	47° 17'	132° 05'	4177	2, 1	1.6	105	5	30
2	51° 30'	131° 14'	3445	0, 0, 0	0	—	—	25
3	53° 09'	130° 32'	3965	0	0	—	—	10
4	54° 42'	129° 43'	4295	0, 6, 3	3.5	270	11	55
5	54° 50'	129° 38'	4665	4, 4, 4	3.9	315	20	35
6	60° 01'	127° 53'	4619	8, 7	7.6	225	3	25
7	54° 43'	127° 58'	4413	8, 9, 7	7.9	070	62*	33
8	50° 32'	-128° 02'	2996	0	0	—	—	20
9	98° 59'	128° 07'	3924	3, 4	3.6	275	19	35
10	45° 06'	128° 02'	5475	0, 0	0	—	—	29
11	38° 08'	127° 59'	5446	0, 0, 0	0	—	—	22
13	34° 50'	124° 25'	4219	5, 5, 5, 5, 5	5.0	025	20	49
15	42° 59'	117° 02'	4734	0, 0, 0, 0	0	—	—	46
16	46° 06'	117° 03'	4120	7, 5, 7, 5, 10	6.8	080	33	48
18	56° 06'	119° 58'	4465	8, 9, 8, 8	8.2	165	—	52
19	47° 30'	128° 00'	4143	7, 7, 7	6.9	240	35	42
20	43° 24'	129° 25'	5041	0, 0, 0, 0	0	—	—	49

* Large difference between two compasses; camera compass used.

Table 1 (cont.)

Eltanin 37

Station	Latitude S	Longitude E	Corrected Depth (m)	Measured Velocities on Successive Hits (cm/sec)	Mean Velocity (cm/sec)	Direction	Difference between Two Compasses	Duration (minutes)
1	65°55'	144°57'	2013	>15	>15	191	—*	5
4	65°17'	140°50'	2196	11, 7, 8	8.8	153	—*	15
5	65°54'	139°01'	622	7	7.1	106	—*	5
6	65°55'	138°54'	684	>15, 11, >15, >15	>15	170	—*	23
7	65°14'	137°51'	2238	6, 4, 5, 6	5.2	305	27	25
8	64°32'	138°03'	3056	3, 5, 6, 6, 8	5.8	353	97**	34
9	64°04'	138°03'	3440	>15, >15, >15, >15, >15, >15	>15	185	38	34
10	64°05'	135°33'	3367	2, 7, 6, 6	5.6	267	18	23
11	64°41'	132°58'	1276	3, 3, 2, 4, 4	3.1	135	-42	45
12	64°02'	130°10'	3160	9, 11, 11, 12, 11	10.7	002	—*	34
13	64°59'	127°17'	2150	>15	>15	279	-10	5
14	63°58'	127°28'	3843	10, 14, 14, 11	12.1	187	-25	22
15	63°02'	127°02'	4127	6, 6, 7, 9	7.1	345	-17	30
16	62°04'	126°57'	4236	4, 5	4.3	192	2	24
17	60°59'	126°22'	4392	3, 0, 0	0	—	—	23
18	60°01'	126°06'	4467	>15, 11, 9, 9	11.0	000	-8	35
19	58°02'	125°36'	4575	4, 4, 6, 4, 5	4.5	073	-32	32
20	56°03'	124°58'	4612	1, 2, 2, 2, 2	1.8	227	-8	30
21	50°56'	125°04'	4000	11, 5, 9, 10, 7	8.2	277	4	37
22	50°59'	122°21'	4209	4, 3, 2, 2, 2	2.7	052	13	35
23	49°54'	120°20'	3916	0	0	—	—	5
24	48°49'	123°09'	4072	5	5.0	022	-5	10

* No camera compass reading.
** Large difference between two compasses; camera compass used.

mean patterns of circulation which may be compared with those derived indirectly on the basis of temperature, salinity and oxygen distributions. The highest velocities (more than 15 cm/sec) occur near ridges or areas of high relief (Tasman and Macquarie ridges) and along the continental slope of Antarctica. A significant increase in turbulence undoubtedly occurs with such high velocities and it is this turbulence which apparently maintains the particles of the nepheloid layer in suspension.

In the South Indian Basin, the mean flow pattern is clockwise. This is apparent when the velocities are averaged over separate parts of the basin (Figure 1). According to II, the salinity distribution indicates dense Ross Sea water spilling into the South Indian Basin along the continental margin at about 140° E. This could be the explanation of the high velocities observed along the margin here. Clockwise circulation around the South Indian Basin would be expected for geostrophic bottom currents circulating thus with the sloping basin boundary on the left.

The Southeast Pacific Basin results are more difficult to explain. The line of measurements made along 120° W shows a pattern of westward flow across the whole basin. The bottom potential temperature distribution (Gordon, 1966) shows a major west to east tongue of cold oxygen-rich water, colder and more oxygen-rich toward the Ross Sea. Besides this same broad tongue across the central part of the basin, Hollister and Heezen (1967) interpreted the data with a narrow cold tongue of water along the south side of the basin, warming to the west. This interpretation indicates a source somewhere to the east, possibly Weddell Sea bottom water. The current meter stations south of 65° S along this section might be explained as part of such a westward flowing mass of Weddell Sea water, but the westward flow across the rest of the basin seems to grossly contradict the potential temperature and O_2 distribution. The best explanation seems to be that the westward flow measured across the central and northern part of the basin is opposite to the mean flow but is not persistent enough in time to cause observable distortion of the isotherms.

The time required for a current of 8 cm/sec to cause an observable change (e.g. a change of 0.05 °C) would be about four months assuming the isotherms (as given by Gordon, 1966) would simply be shifted an amount equal to the displacement of water. The stations of this section between 55° and 65° S were taken over a time span of ten days, which might be considered a lower limit of the period of duration of this irregular flow. Whatever the cause of this westward flow which we believe to be opposite to the mean flow eastward, it demonstrates the necessity of obtaining long period time series measurements in order to test adequately conclusions about bottom circulation based on distributions of properties.

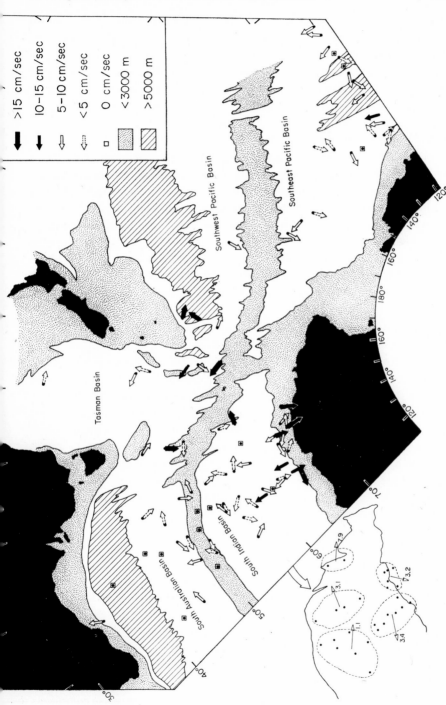

Figure 1 Bottom current measurements. The South Indian Basin measurements have been averaged in geographic groups and the results shown in the lower left corner, with the vectorial average direction and magnitude in cm/sec shown. Bathymetry from Heezen (in preparation)

NEPHELOMETER RESULTS

Four examples of nephelometer profiles, two in each of the southern basins, are shown in Figure 2. The curves are similar to those recorded in other areas of vigorous bottom water circulation, such as the western trough of the Atlantic Ocean (Ewing and Thorndike, 1965; Ewing, *et al.*, in press; Eittreim, *et al.*, 1969; Eittreim and Ewing, this volume). The curves show a high in light scattering intensity near the surface, a gradual decrease with depth to a minimum usually in the lower part of the water column and then an increase in the bottom waters. This increase in scattering in the bottom waters is the nepheloid layer which will be of primary concern here. The gradual decrease in scattering with depth in the upper 2 or 3 kilometers we believe is caused by organic material associated with the near-surface euphotic zone. Profiles of organic particulate matter concentration at a number of stations in the North Atlantic show a similar, although more erratic, decrease with depth (Gordon, 1970a,b).

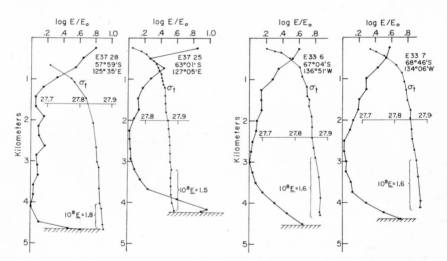

Figure 2   Light scattering intensity and density (sigma-*t*) profiles for two stations in the South Indian Basin (left-hand side) and Southeast Pacific Basin (right-hand side). Values of stability, *E* (Neuman and Pierson, 1966), were calculated for hydrographic data points in the nepheloid layer. The light scattering units are log exposure (*E*) of nephelometer film relative to exposure in the clearest water (E_0)

Along with the vertical profiles of light scattering, sigma-*t* profiles are also shown. These are calculated from the temperature and salinity profiles for the same stations taken from the *Eltanin* data reports which are in preparation by Jacobs and others. The intensity of vertical turbulence is closely linked to the density stratification, since buyancy forces must be overcome to displace water vertically in a column with stable density stratification. Stability, E, is a measure of the density stratification and this was computed from the hydrographic data in the nepheloid layer at these four stations (Figure 2). The particles of the nepheloid layer are maintained in suspension by upward eddy diffusion which balances (or partially balances) the downward gravitational flux. For a given gravitational flux, large vertical gradients in particulate matter should reflect small eddy diffusion coefficients and conversely, small gradients in particulate matter should reflect large eddy diffusion coefficients or the presence of a well-mixed water layer. On this basis, the nepheloid layer at station E37 N28 (Figure 2) with the strongest gradient of particulate matter represents a zone with the weakest eddy diffusion among the four stations shown. This is in accord with the highest value of stability of any of the four profiles. Note that although the density stratification is weaker in the bottom waters relative to other depth regions, these waters do maintain positive stratification. In contrast, the bottom nepheloid layer of the North Pacific occupies a region of neutral stability (Ewing and Connary, 1970).

The exposure of film on the nephelometer record is proportional to the intensity of light scattering. The measured parameter of the negative film is the optical density, which is related to exposure by a log relationship. Hence the units used are log exposure. In Figure 2 the curves are normalized to the clearest water in the water column. This procedure of using the clearest water as a reference is somewhat hazardous, especially near the continent where significant amounts of terrigenous detritus might cloud the water at mid depths.

The intensities at the bottom, contoured in Figure 3, are in the same units and, since the stations between 120° E and 160° E along Antarctica are located near the continental edge, the values shown for these stations might be considered lower limits.

Figure 3 illustrates that the nepheloid layer is mainly confined south of the mid ocean ridge. A notable exception is the region north of the Macquarie Ridge–mid ocean ridge intersection. A northward tonguing of turbid water in this region, a tonguing which is also seen in the potential temperature distribution (see II), indicates a northward passage through the ridge of turbid, cold bottom water into the Tasman Sea. Therefore, with the above exception, the mid ocean ridge acts as an effective northern barrier to the turbid bottom water which apparently has its origin near Antarctica. Since the high intensities do not extend far into the Tasman Basin, it appears that the suspended material is dropped fairly rapidly upon entering the basin. Houtz, *et al.* (in press) show the presence of an isolated pod of acoustically

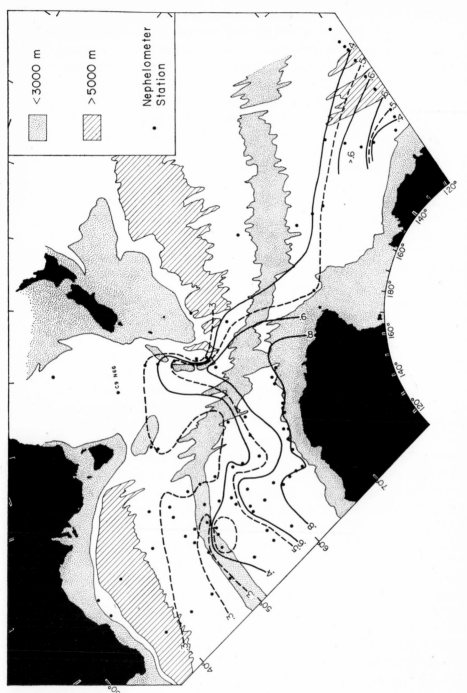

Figure 3 Near bottom light scattering intensities in units of log film exposure relative to exposure in the clearest water. Bathymetry from Heezen (in preparation). Except for the labeled station in the Tasman Sea, *Conrad* cruise 9, station 66, all data are from *Eltanin* cruises

homogeneous sediment centered on about 55° S and 157° E, a location which is approximately in the path of this entering bottom water (seismic profile D-1, Figure 3 in the above reference). This sediment body has a maximum thickness of 600 m and a horizontal extent of approximately 100 km, elongated in a north–south direction. It is far from terrigenous sources and too localized to be related to surface water biologic productivity changes. Fallout of material from a turbid high-velocity bottom current here, at a point where it enters a relatively tranquil basin, would explain the existence of this sediment body.

Two important factors which are necessary to maintain a nepheloid layer are (a) turbulence in the bottom waters generated by high-velocity bottom currents and (b) a source of particulate matter to this turbulent layer either by erosion from the bottom or direct introduction from the continent. In the South Indian Basin, the highest intensities occur adjacent to Antarctica. This is probably a reflection of proximity to source of particulate matter, since high-velocity bottom currents, capable of erosion, are not unique to this area, but are measured at many places all over this region of study (see Figure 1). Notably, velocity measurements of more than 15 cm/sec were made south of the Tasman Ridge and in the Southwest Pacific Basin south of New Zealand, neither of which is in regions of high-intensity nepheloid layer. New bottom water, derived from the shelf and slope of Antarctica, is likely to contain high amounts of particulate matter, both organic material from the euphotic zone and mineral matter derived from the erosion of the Antarctic continental shelf sediments. Therefore, the turbidity of the bottom water may be largely a function of its "youth", the term youth here referring to the time since formation of the Antarctic Bottom Water.

From the data on bottom water oxygen concentrations given in II, (Figure 6) we note that the highest oxygen values occur in the South Indian Basin, where the highest light scattering intensities are also recorded. This correlation is illustrated in Figure 4. The oxygen values are taken from the *Eltanin* data reports of Jacobs and others for all hydrographic stations in this region, at which a tripod lowering was also made. Oxygen is acquired by the water when it is on the continental shelf and in contact with the atmosphere or euphotic zone waters prior to sinking. Its concentration is decreased with time by mixing with overlying water and by organic consumption. Therefore oxygen content is recognized as a good indicator of the youth of bottom waters. A tongue of oxygen maximum also extends from the Ross Sea northward and eastward in a zone occupying the deepest part of the Southeast Pacific (Bellingshausen) Basin. Gordon (1966) attributes this oxygen maximum (which is also a potential temperature minimum) to a source of bottom water in the Ross Sea. As shown in Figure 3, this region is also associated with a light scattering maximum, probably reflecting again the youth of the bottom water.

The thickness of the nepheloid layer in general increases toward the centers of the basins to an average thickness of about 1 km. Figure 5 shows the depth to the clearest water measured by the nephelometer. This depth

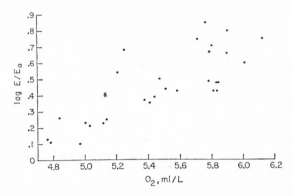

Figure 4   Light scattering versus oxygen content of bottom water samples taken at nephelometer stations in all five basins. Oxygen values are the same as those in Gordon (this volume)

may be considered as the "top" of the nepheloid layer, since generally there is a steady increase in light scattering down from this level (Figure 2). In the South Indian Basin the depth of the clearest water is greatest in the center of the basin, the nepheloid layer thus having a bowl shape. This implies an association with clockwise circulating bottom water. This clockwise circulation is seen in the bottom current measurements (Figure 1) and is implied by the distribution of temperature, salinity and oxygen (see II). The same pattern appears to be present in the South Australian Basin (Figure 5) but the data are too sparse for confidence in the contours of the northeast part, west of Tasmania.

CONCLUSIONS

The great irregularity of current directions observed with short time period (one hour) measurements makes obvious the need for long period time series measurements to deduce mean flow patterns and to better understand the deep water motions. Observations of velocities greater than 15 cm/sec were numerous and were generally in regions of rough topography or along the continental slope of Antarctica. Such velocities, coupled with the shear-producing irregular motions which are observed are probably responsible

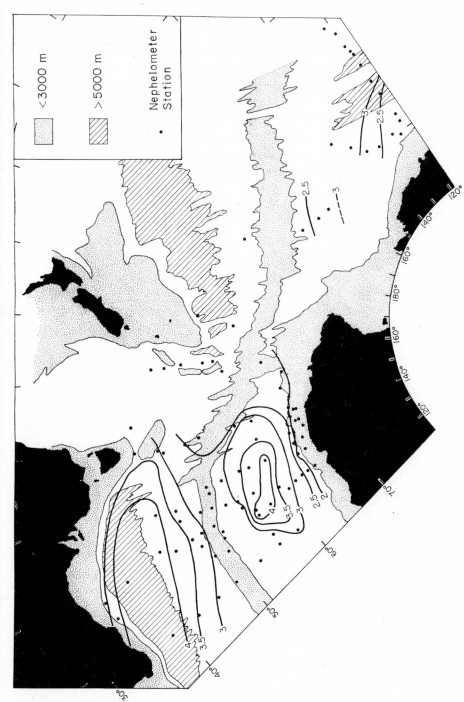

Figure 5 Depth in km to the clearest water in the water column which can be thought of as the "top" of the nepheloid layer

for turbulence in the bottom waters. This turbulence will produce a nepheloid layer when a source of particles to the turbulent layer exists.

A mean flow pattern of clockwise circulation is seen in the South Indian Basin. Westward flow which contradicts the mean flow direction deduced from temperature and oxygen distribution is observed in the deep central Bellingshausen Basin.

The nepheloid layer, which exhibits the most intense light scattering at stations adjacent to Antarctica, represents a zone in which there is an increase in intensity of scattering of a factor of six relative to the clearest water. This is somewhat less than the factor of 10 increase found in the North American Basin nepheloid layer.

The high intensity nepheloid layer lies south of the mid ocean ridge except at its intersection with the Macquarie and Tasman ridges. In the latter region, a northward tongue of turbid, cold water extends into the Tasman Sea.

Particulate matter is associated with newly formed Antarctic Bottom Water (AABW) which has acquired particles mainly from the near-surface waters of the Antarctic shelf. This conclusion is supported by (a) the occurrence of highest light scattering intensities near Antarctica where sources of AABW are likely, and (b) the correlation of high intensities with high oxygen content. Erosion of the continental slope and deep basin floor may be a secondary source of particles to the nepheloid layer.

The nepheloid layer has a bowl-shaped geometry in the South Indian and possibly South Australian basins indicating an association of the layer with clockwise circulating bottom waters.

Although turbulence is required to maintain the particles in suspension, this turbulence is limited by the density stratification observed. The density profiles at four stations show positive stability in the nepheloid region although the stability is weaker here than elsewhere in the water column.

Acknowledgments

R. Houtz, K. Hunkins and J. Nafe critically reviewed the manuscript. Discussions of the data with S. Jacobs were most helpful. This work was supported by National Science Foundation grants GV 19032 and GA 1615.

References

Deacon, G.E.R., The hydrology of the Southern Ocean, *Discovery Rep.* **15**, 1–124, 1937.
Eittreim, S. and M. Ewing, Suspended particulate matter in the deep waters of the North American Basin, this volume.
Eittreim, S., M. Ewing, and E.M. Thorndike, Suspended matter along the continental margin of the North American Basin, *Deep-Sea Res.*, **16**, 613–624, 1969.
Ewing, M. and S. Connary, Nepheloid layer in the North Pacific, *Geol. Soc. Am. Mem.*, ed. J. Hays, in press.

Ewing, M., S. Eittreim, J. Ewing, and X. Le Pichon, Sediment transport and distribution in the Argentine Basin: Part 3, Nepheloid layer and processes of sedimentation, *Physics and Chemistry of the Earth*, **8**, Pergamon Press, in press.

Ewing, M. and E. M. Thorndike, Suspended matter in deep ocean water, *Science*, **147**, 1291–1294, 1965.

Gordon, A. L., Potential temperature, oxygen and circulation of bottom water in the Southern Ocean, *Deep-Sea Res.*, **13**, 1125–1138, 1966.

Gordon, A. L., Recent physical oceanographic studies of Antarctic waters. In: *Antarctic Research*, ed. L. Quam, AAAS, Washington, D.C., in press.

Gordon, A. L., Spreading of Antarctic bottom waters, II, this volume.

Gordon, D. C., A microscopic study of organic particles in the North Atlantic Ocean, *Deep-Sea Res.*, **17**, 175–185, 1970a.

Gordon, D. C., Some studies on the distribution and composition of particulate organic carbon in the North Atlantic Ocean, *Deep-Sea Res.*, **17**, 233–243, 1970b.

Heezen, B. C., Antartic bathymetry, *Antarctic Map Folio Series*, ed. V. Bushnell, in preparation.

Hollister, C. D. and B. C. Heezen, The floor of the Bellingshausen Sea. In: *Deep-Sea Photography*, ed. J. B. Hersey, Johns Hopkins University Press, 177–189, 1967.

Houtz, R. E., J. Ewing, and R. Embley, Profiler data from the Macquarie Ridge area, *Antarctic Research Series*, Am. Geophys. Union Monogr., in press.

Jacobs, S. S., A. F. Amos, and P. Bruchhausen, Ross Sea oceanography and Antarctic Bottom Water formation, *Deep-Sea Res.*, in press.

Neuman, G. and W. J. Pierson, *Principles of Physical Oceanography*, Prentice-Hall, New York, 545 pp., 1966.

Reid, J. L. and W. D. Nowlin, Transport of water through Drake Passage, *Deep-Sea Res.*, in press.

Thorndike, E. M. and M. Ewing, Photographic nephelometers for the deep sea. In: *Deep-Sea Photography*, ed. J. B. Hersey, Johns Hopkins University Press, 113–116, 1967.

Thorndike, E. M. and M. Ewing, Photographic determination of ocean-bottom current velocity, *Mar. Tech. Soc. J.*, **3**, 45–50, 1969.

Paper 15

Geologic Effects of Ocean Bottom Currents: Western North Atlantic*

CHARLES D. HOLLISTER

Woods Hole Oceanographic Institution
Woods Hole, Massachusetts 02543

BRUCE C. HEEZEN

Lamont-Doherty Geological Observatory, Columbia University
Palisades, New York 10964

Abstract Bottom photographs, sediment cores and high resolution echograms from the Atlantic continental margin of North America reveal distinctive features created by the Western Boundary Undercurrent, a deep current associated with the thermohaline circulation of the Atlantic.

Current lineations observed in oriented bottom photographs taken on the continental rise show features indicative of significant sediment transport and deposition parallel to bathymetric contours. Bottom current directions inferred from the orientation of current lineations correspond to the flow direction of the Western Boundary Undercurrent predicted long ago by the pioneering work of Georg Wüst.

Sediment cores reveal a distribution of distinctive brick-red clay that has apparently been transported from the Cabot Strait southwesterly to the Blake-Bahama Outer Ridge.

Massive muddy sands are typically found on abyssal plains. The graywacke-type sands from the abyssal plains may exceed 50 cm in thickness, and as a rule, rarely exceed 50 per 10 m of core. The continental rise is, on the other hand, underlain by sediment containing as many as 500, thin (< 1 cm), well-sorted silt lamina per 10 m. These silts are relatively clay-free and usually contain cross-beds accentuated by heavy mineral placers.

The continental rise appears to have been constructed by: 1) material injected laterally into low velocity (< 30 cm/sec) contour-following bottom currents by high velocity (> 100 cm/sec) turbidity currents flowing downslope; and 2) continental material transported in suspension through the water column and subsequently mixed with pelagic components to form hemipelagic sediment.

The very largest and most competent turbidity currents reach the abyssal plains where

* Woods Hole Oceanographic Institution Contribution Number 2620.
Lamont-Doherty Geological Observatory Contribution Number 1765.

37

they deposit relatively thick beds of muddy, graded coarse sand, silt and clay known as *Turbidites*. The well-sorted fine sand, silt and clay laminations deposited by contour currents flowing on the continental rise are known as *Contourites*.

INTRODUCTION

The high probability that bottom water circulation plays a significant role in the transportation of abyssal sediments was recognized long ago by Professor Georg Wüst (1936). However, geologists were slow to grasp the importance of the 1925–27 *Meteor* results and as late as 1958, Professor Wüst was moved to publish a paper in a geological journal calling the attention of geologists to the probable geologic effects of the deep currents inferred from the detailed analysis of the *Meteor* data (Wüst, 1958). This paper apparently escaped general notice and it was not until Professor Wüst exerted his personal influence on various workers, including the authors, that the study of the geologic effects of deep-sea bottom currents began.

Wüst's efforts stimulated studies of bottom current evidences such as those recorded in photographs and preserved in deep-sea cores. Such studies revealed compelling evidence that bottom currents associated with global circulation have an important effect on sediment distribution in the abyss (Heezen and Hollister, 1963, 1964).

Ripples and lineations produced by bottom currents have now been photographed in abyssal depth beneath many deep current systems (Heezen and Johnson, 1964; Heezen, Schneider and Pilkey, 1966; Hollister and Heezen, 1967; Hollister and Elder, 1969; Heezen and Johnson, 1969).

Further investigations using oriented bottom photographs, precision echograms and long piston cores demonstrated the close relationships between bottom circulation patterns and the shaping of large sediment bodies in the deep sea (Heezen, Hollister and Ruddiman, 1966; Hollister, 1967; Schneider, Fox, Hollister, Needham and Heezen, 1967; Jones, Ewing, Ewing and Eittrem, 1970).

Acknowledgements

Informative discussions with J.R.Conolly, S.Eittrem, K.O.Emery, D.W. Folger, P.J.Fox, A.Gordon, L.King, R.Mead, D.H.Needham, J.Reid, W.B.F.Ryan, J.Schlee, H.Stommel, W.A. von Arx, and L.V.Worthington are gratefully acknowledged. The Bedford Institute of Oceanography, American Telephone and Telegraph Company, Cable and Wireless, Ltd., Duke University Marine Laboratory and Narragansett Marine Laboratory provided ships' time for sea-floor photography. Most of this work was carried out with the support of the Office of Naval Research, The National

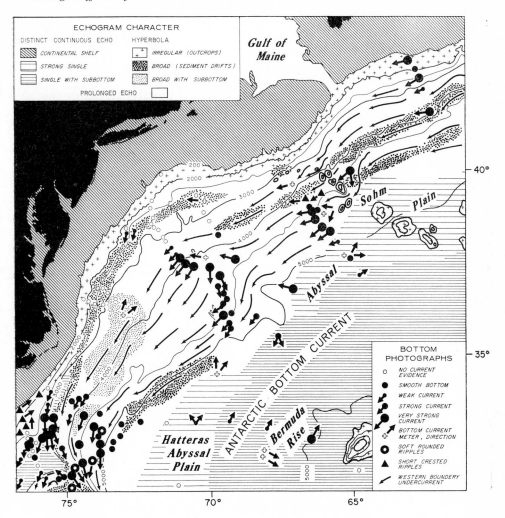

Figure 1 Direction and strength of Western Boundary Undercurrent inferred from bottom photographs. Strong coherent echos return from continental shelf and abyssal plains. On the continental rise zones of prolonged echoes and hyperbola trend parallel to the regional contours. Above 3500 meters, tranquil, current-free bottom is usually seen but below 3500 meters, swift contour currents appear to be transporting sediment towards the south. Photographs on the Hatteras Abyssal Plain show tranquil bottom or variable week northerly-flowing currents, while swifter northerly flow of the Antarctic Bottom Current is observed on the Western Bermuda Rise. The Gulf Stream, not shown, apparently reaches the bottom at least occasionally in abyssal depths. Photographs, taken beneath the surface axis, often show evidence of a north to northeasterly flow. (Contours in meters)

Science Foundation and the Bell Telephone Laboratories. Support from the United States Geological Survey (Grant, U.S.G.S. 12109) and the Office of Naval Research (Grant, ONR 241-9) is gratefully acknowledged.

CURRENT EVIDENCE FROM BOTTOM PHOTOGRAPHS

Several hundred-thousand photographs have been taken of the deep-sea floor sampling 5 to 25 m² rectangles of virtually every known environment (Heezen and Hollister, 1971). Throughout most of the world ocean the dominant features seen in photographs are traces of animal life. Bottom photographs taken in the western North Atlantic frequently reveal an anomalously smooth bottom and others record ripple marks and current lineations (Figures 1, 2).

Currents interacting with bottom sediments can produce visible effects such as current lineations, scour marks, bare rock and coarse debris. Symmetrical or asymmetrical, transverse or longitudinal, straight-crested, curving or irregularly-crested ripple marks are observed in bottom photographs. Wavelengths range from centimeters to meters and occasionally to kilometers. Amplitudes in turn range from one centimeter to many meters. Scour marks around nodules, rocks, and other solid objects are often observed. In regions of very high velocities, rocks are concentrated in depressions ("rock nests" of Hollister and Heezen, 1967).

The gentlest current lineations often consist of subtle features. Currents are often indicated simply by the absence of burrow mounds and tracks or by mounds and burrows deformed or partially destroyed by currents. Slightly elongated burrows, a noticeable fabric or a certain parallelism of small (<2 cm) sediment tails behind feces and other resistant objects is frequently all that is observed.

Current-smoothed bottom contrasts markedly with the greater and more random irregularities of the "normal" ocean floor (Figure 2). Lineations often become more marked and one can then discern definite individual drifts. Strong elongations of sediment behind burrows and other objects and appreciable sediment accumulation in tails a centimeter or more in thickness is revealed in photographs by a deepening of the shadows. Currents of higher velocity create scour lineations which are more marked. Under the influence of higher velocities moats as well as tails develop around solid objects.

Ripples form the smallest limit of a continuum of sea floor sediment waves which range in size from centimeters to kilometers. Thousand-meter-long sediment waves cannot be photographed in deep water but underwater photographs have revealed smaller ripple marks in depths to 7500 m. Although few ripple marks and scour marks are seen in the greatest depths there appears to be no demonstrable relationship between the depth of water and the size and shape of those that have been seen. Short-crested ripple marks readily develop in sand-size material and the larger, rounded forms seem to have been created in fine oozes and clays.

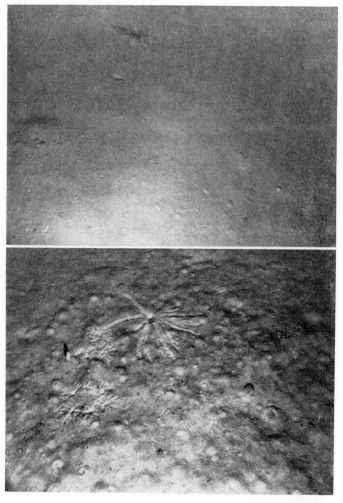

Figure 2 Smoothed muddy bottom on the Bermuda Rise and soft burrowed and tracked bottom on the west flank of the Mid-Atlantic Ridge. These two photographs serve to illustrate the contrast in bottom character between the current-smoothed Bermuda Rise (mostly deeper than 4500 m) and the tranquil bottom typically seen in photographs from the flanks of the Mid-Atlantic Ridge (mostly shallower than 4500 m). (Top, A. *Snellius*-I6; 5070 m; 33° 59′ N, 65° 00′ W; Bottom, B. *Snellius*-N2; 3669 m; 48° 51′ N, 34° 10′ W)

Lineations become difficult to interpret when all loose sediment including sand and gravel are removed. Under such strong currents there is no sure way to ascertain either the current magnitude or direction. Deep-sea photographs have usually been taken without the use of a compass and consequently directional features cannot be oriented.

The critical traction transport velocities for deep sea sediment particles, if determined, could provide estimates of minimum velocities required to

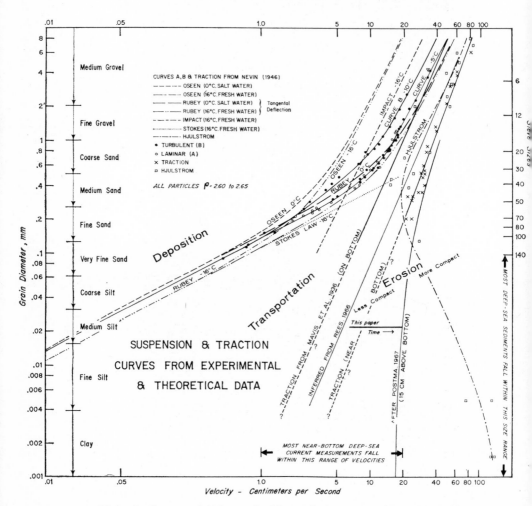

Figure 3 Current velocities required for erosion, transportation and deposition of stream bed materials from equations determined theoretically and empirically. Recent experiments (Rees, 1966; and Southard *et al.*, 1971) indicate that traction transport of medium silt-sized material can occur at velocities above about 7 cm/sec. (Redrawn from Heezen and Hollister, 1964)

form ripples, scour marks, current lineations and other current features. Traction transport velocities have been determined experimentally, but stream conditions rather than sea-floor conditions have been approximated (Figure 3). Flume experiments show that well-sorted fine silt grains are eroded from a plane bed at velocities of about 5 cm/sec; and after excess fine silt was added, asymmetrical ripples have been observed to develop at velocities between 8 and 13 cm/sec (Rees, 1966). According to the stream data of Hjulstrom (1935) obtained on less well-sorted natural sediments, velocities of about 80 cm/sec would be required for erosion of material having a similar average grain size as that used by Rees.

In order to initiate motion, particles of about 0.02 mm might only require a current velocity of 4 cm/sec according to one extrapolation of the traction curve (Mavis *et al.*, 1935), 7 cm/sec according to recent experiments of erosion in fine carbonate ooze (Southard *et al.*, 1971), and 50 cm/sec (about 1 knot) according to another interpretation (Hjulstrom, 1935). Particles of 0.2 mm would require for erosion a velocity of 10 cm/sec according to the Mavis curve and 20 cm/sec according to Hjulstrom. If deep-sea clays are less cohesive (due to flocculation effects of sea water) than the river bottom clays used in most traction experiments, velocities required to transport deep-sea silts and clays may be as close to the extrapolated values of Mavis as recent experiments seem to indicate.

Critical traction velocities as well as the velocities needed for transport in water containing appreciable suspended sediment have been studied (Kuenen, 1965). Tank experiments with relatively dense suspensions (>1 gm/cc) of clay indicate that: 1) At velocities of 30–50 cm/sec no deposition occurs; 2) Below 10 cm/sec all clay ultimately settles out; 3) Between 10 and 50 cm/sec the capacity is dependent on velocity; 4) For overloaded currents percentage fallout per unit of time is dependent on velocity and independent of sediment concentration; 5) On smooth bottom clay was not brought into suspension below velocities of about 20 cm/sec but erosion occurred on a slightly irregular bottom at 12 cm/sec; 6) Particles that come in contact with clay bottom stick to it; and 7) Velocities two or three times those which allow deposition are required to erode these particles.

BOTTOM CURRENTS ON THE CONTINENTAL MARGIN

Photographs taken on the continental slope south of Cape Cod reveal a predominantly muddy bottom with many tracks, trails, and burrow mounds as well as abundant bottom dwelling organisms such as star fish, brittle stars, sea pens and sea spiders (Figure 4). Evidence of currents in the form of lineations (Emery and Owen, 1967) and ripple marks (Elmendorf and Heezen,

1957) on the continental slope of eastern North America are noted in about $\frac{1}{4}$ of the available camera stations.

At one location at 606 m off Nova Scotia ripples and scour marks were seen around scattered rocks on a sandy bottom. Current direction inferred from asymmetrical ripples as well as the bending direction of attached organisms indicates a southwesterly current. At another station near the continental rise—continental slope boundary off Nova Scotia (1561 meters) the bottom appeared smoothed and some of the over 400 photographs taken at this locality showed poorly developed current lineations.

Figure 4 Soft muddy bottom continental slope off New York. Most photographs taken on the continental slope off eastern North America show some effects of bottom currents; however, they also reveal abundant mounds, holes and other bioturbations. (Photo courtesy of G. T. Rowe). *Atlantis* II–52 (1513); 1850 m; 39° 12′ N, 71° 52′ W

Volkmann (1962) made two series of direct current measurements (in 1959 and 1960) in the slope water south of Cape Cod along 71° and 68° west longitude. These measurements, taken approximately 300 to 500 miles west of the camera stations mentioned above showed a westward flow of water with velocities greater than 20 cm/sec from the surface to depths of more

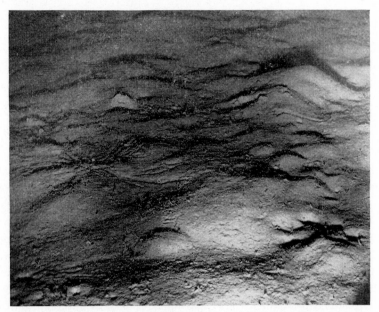

Figure 5 Tranquil bottom conditions exist on the modern surface of the upper continental rise. Here mounds and holes are preserved and current smoothing is usually absent. (*Gosnold* 199; 3026 m; 38°30′ N, 70°36′ W). Photo courtesy of D. W. Folger

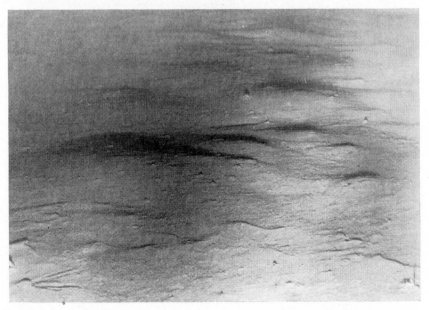

Figure 6 Burrows, mounds and holes are smoothed and lineated at the base of the lower continental rise. Currents here are flowing parallel to local contours. (*Gosnold*–202; 5010 m; 36°47′ N, 67°05′ W). Photo courtesy of D. W. Folger

than 3200 m. He concluded that these data "... support the existence of a deep western boundary current flowing along the coast of the United States" (Volkman, 1962). Sediment put into suspension by the abundant vagile benthic organisms would probably be swept away if current velocities were on the order of 10 cm/sec. Indeed, the absence of appreciable sediment cover on the slope in this region as determined by seismic profiles (Knott and Hoskins, 1968) suggests that this environment is one of non-deposition or erosion.

Along the base of the continental slope off Beaufort, North Carolina, and along the Blake-Bahama Outer Ridge, current lineations consisting of streamers of sediment deposited in the lee of burrow mounds and other objects on the bottom have been frequently observed. On the eastern side of the Outer Ridge between 3000 and 5000 m and along the base of the continental slope off North Carolina abundant lineations indicate a southerly current flowing parallel to the local contours (Figure 1). Where current lineations are abundant, the bottom is smooth and remarkably free of benthic life and the water appears muddy. Camera cores taken at each station contain foraminiferal gray and light brown silty clay. Silt and fine

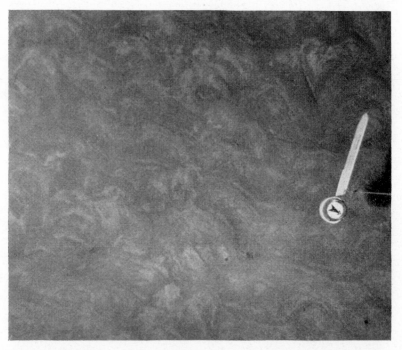

Figure 7   Short-crested ripples beneath the Western Boundary Undercurrent indicate strong bottom flow in a westerly direction in 4410 m depth. (*Trident*–052 (3); 39° 22′ N, 66° 04′ W), (Photo courtesy of H. Zimmerman)

sand-sized foraminiferal tests range from 5% to 15%, and clastic silt consti-
tuted approximately 15% of the sediment. Velocities required for movement
of material of this size are probably less than 15 cm/sec. Near-bottom current
velocities of up to 18 cm/sec have been measured in the southerly-flowing
Western Boundary Undercurrent (Swallow and Worthington, 1961) less than
30 miles from an oriented camera station showing well-developed, southerly-
oriented, current lineations. In contrast to photographs taken on the Blake-
Bahama Outer Ridge, photographs from the convex-up upper continental
rise between the Outer Ridge and Georges Bank show a tranquil, muddy
bottom with many burrow mounds and holes (Figure 5). Photographs in
deeper water on the relatively flat lower continental rise in this region where
the rise can be divided into two distinct portions show very well-developed
current lineations indicating a bottom current flowing in a southwesterly
direction parallel to local contours (Figures 6, 7).

Off Nova Scotia, where the entire rise is essentially a smooth sedimentary
apron built against the continental slope very strong current evidence is
observed (Figure 8). Soft rounded asymmetrical ripples (crest orientation
N–S) were seen in over half of the 100 photographs taken at this station. The
orientation of the well-developed current lineations and the direction of sediment dispersion (caused by the compass hitting the sea floor) indicated a
current flow toward the west.

Unoriented photographs taken on the continental rise (3910 m) north of
the Grand Banks show soft rounded ripples, lineations (Figure 9), and a
smoothed muddy bottom. Photographs at 2 locations on the gently sloping
continental rise southwest of Cape Farewell, Greenland, show preferential
deposition of material on the downcurrent side of burrow mounds in addition
to well-developed current lineations. These depositional forms, though not
as dramatic as ripples, are sufficient to show the effects of appreciable currents on the bottom. A northwesterly-flowing current of 10 cm/sec was
measured 400 m above the bottom in this area (Swallow and Worthington,
1969).

Nearly all of the approximately 5000 bottom photographs examined from
the continental slope, continental rise and outer ridges along the western
North Atlantic from Greenland (Heezen and Hollister, 1964) to the West
Indies (Ewing and Mouzo, 1968; McCoy, 1968) show smoothing, well-
developed current lineations, ripples, and other current effects. It seems
reasonable to surmise that these bottom current effects on the continental
margin are formed by the Western Boundary Undercurrent that flows along
the continental margin as part of a steady-state thermohaline circulation of
the North Atlantic.

Recent work (Eittrem *et al.*, 1969) indicates that the Western Boundary

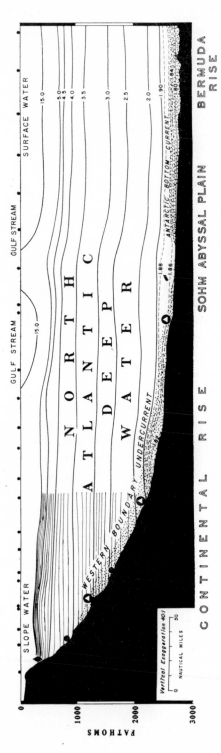

Figure 8 Sediment, bottom photographs and potential temperature projected (along contours) to a topographic profile off Nova Scotia. In this region the continental rise cannot be divided topographically into an upper and lower portion. Recent sediment is thickest near the base of the comparatively short continental slope and thins to less than 5 cm near the continental rise-abyssal plain boundary (Core V18-375). The configuration of near bottom isotherms together with current evidence seen in bottom photographs suggest that the strongest current probably occurs between Cores V16-213 and V18-375. In contrast to sediment on the continental shelf, slope and abyssal plain, sediment on the continental rise consists of numerous thin silt laminations with heavy mineral placered cross beds in addition to abundant red-colored lutite. Sand beds over 50 cm thick in the deep sea are restricted to the abyssal plain where there is little or no evidence of bottom currents

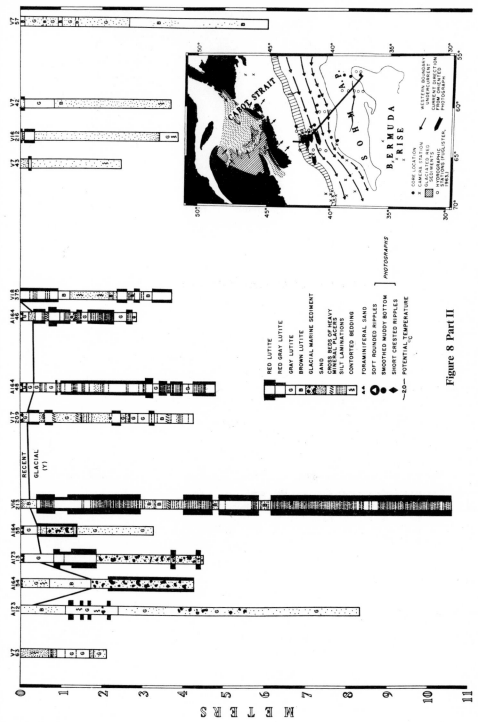

Figure 8 Part II

Figure 9 Soft, rounded ripples, scour moats and lineations beneath the Western Boundary Undercurrent on the continental rise off Labrador. (*Snellius*–HO(5); 3970 m; 51°53′ N, 46°38′ W)

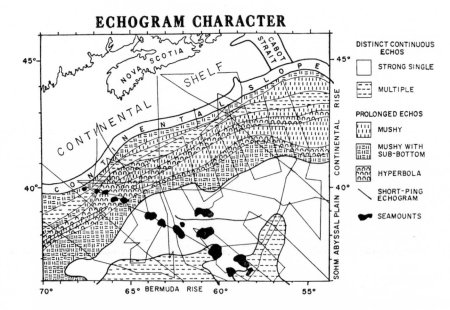

Figure 10 Echogram character on the continental margin and Sohm Abyssal Plain off Nova Scotia. Zones of similar echo reflectivity generally follow bathymetric contours and can easily be correlated from profile to profile. The zones of poorest echo reflectivity also run parallel to near-bottom isotherms, thus suggesting a relationship between echo reflectivity and water mass characteristics. Only the most recent echograms utilizing short outgoing pulse lengths (less than 2–5 milliseconds) were used

Undercurrent is presently carrying significant amounts of suspended material in a layer up to 1 km thick along the bottom on the continental rise and Blake-Bahama Outer Ridge off eastern North America. These observations further substantiate the view that the Western Boundary Undercurrent is a significant agent of sediment transportation along the continental rise. Distinctive echo types appear to follow bathymetric contours (Figure 10). The persistence of such zones at relatively restricted depth ranges from one profile to another suggests that the process responsible for the observed difference of echogram character is in some way related to water depth.

The zone of poorest reflectivity occurs along the middle and lower portion of the continental rise (between 3800 and 4900 m). It is within this depth range that the densest (σ_t 27.89) bottom water occurs. Here the current must flow approximately parallel to bathymetric contours. If the measured bottom current velocities of 10 to 25 cm/sec are competent to transport the sediment type (clay, silty-clay and silt) generally found on the continental rise (Figure 3), then one might conclude that the reflectivity of this region is probably related to the distribution of small surface forms such as ripples as well as the density

Table 1 Primary sedimentary structures of deep-sea sands and silts from the Western Atlantic

Sedimentary Structures	Abyssal Plains	Submarine Canyons	Continental Rise
Massive sand beds thicker than 50 cm	Very common (except near seaward margin)	Common	Rare
Cross beds of heavy mineral placers	Rare (except near continental rise boundary)	Rare	Very common
Cross beds of clay concentrations	Common (good examples have been found in Tyrrhenian, Colombia, and Puerto Rico Trench Abyssal Plains)	Infrequent	None
Laminations of heavy mineral placers in sand or silt bed	Rare	Rare	Very common
Laminations of clay concentrations in sand or silt beds	Very common	Very common	Rare
Alternating thin beds and lamina of silt and clay	Common on seaward margin only	Rare	Very common
Graded silts and sands	Common	Common	Common

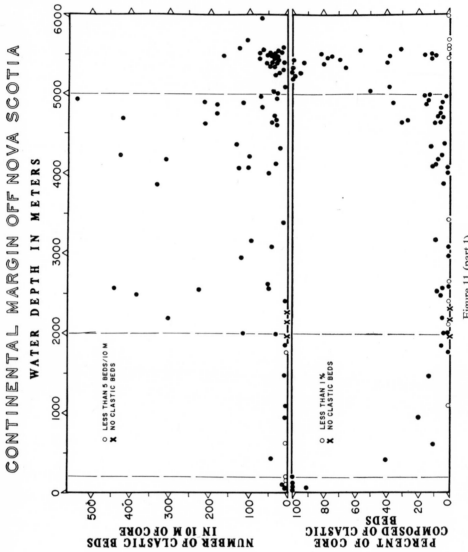

Figure 11 (part 1)

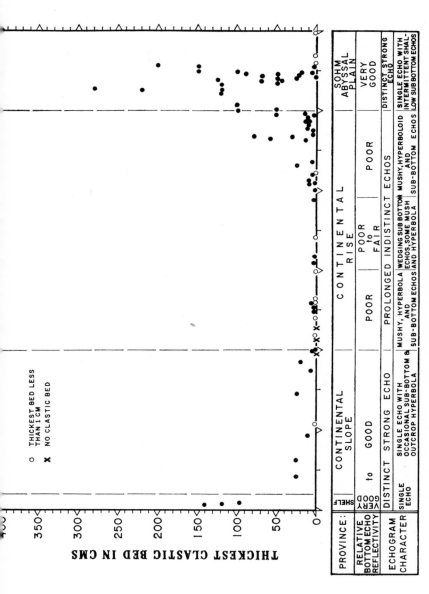

Figure 11 Distribution of clastic beds in cores and bottom echo reflectivity on the continental margin and abyssal plain off Nova Scotia. The highest number of thin clastic beds occurs on the continental rise where bottom echo reflectivity is poorest (bottom echoes are hyperbolic and mushy). On the continental shelf and abyssal plain, where the beds are thick and few in number, bottom echo reflectivity is comparatively good.—Each bed or lamination of clastic material, even if only a few grains thick, was counted in each core. Some layers in a few dried cores were badly disturbed or nearly completely destroyed and in such cases the number of beds in the questionable zone was estimated by interpolating from the average bed counts above and below

and stratification of the upper few meters of sediment (Figure 11) which is controlled by a bottom current—the southerly-flowing Western Boundary Undercurrent.

SEDIMENTS OF THE CONTINENTAL RISE

The continental rise is composed of thick (up to 2 km) seaward-thinning, wedge-shaped aprons of sediment that lap against the base of the continental slope (Knott and Hoskins, 1968; Emery *et al.*, 1970). It is covered with fine-grained gray, brown or red hemipelagic silty clay interbedded with numerous thin silt laminations (Figure 8). Carbonate content generally ranges between 20% and 30%. However, some samples on the continental rise off New

Figure 12 Cross laminations accentuated by concentrations of heavy mineral grains in clastic silt beds on the Newfoundland Continental Rise (× 1, photograph by direct projection of thin section). Virtually all of the comparatively thick (1–5 cm) sand and silt beds from the continental rise are cross laminated and cross laminations can occur anywhere in the bed. The vast majority of cross laminations found on the continental rise are composed of mud-free layers (alternately rich and poor in heavy minerals). Magnetite, hornblende, garnet and pyroxenes are the predominant heavy minerals. These cross laminated sequences have little or no interstitial clay matrix, and voids are filled with the epoxy-resin impregnating compound. This sequence shows "criss-cross" bedding as the strata are inclined obliquely to one another and to the horizontal bedding plane. (*Atlantis*, 157-2 (90)).

(After Heezen and Hollister, 1971)

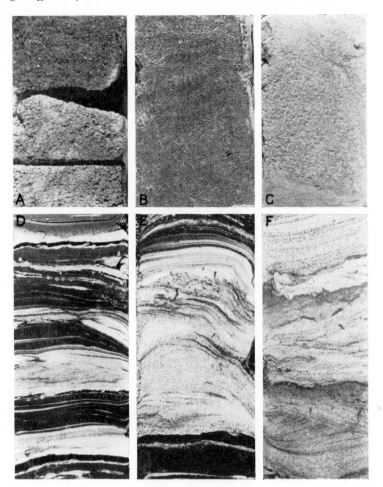

Figure 13 Graded, massive, muddy sands from the Sohm Abyssal Plain (top) and graded highly stratified, cross-laminated clean fine sand and silt beds from the continental rise off New England and Labrador (bottom). (All photos ×½). The cross laminations (bottom) are accentuated by concentration (placers) of heavy minerals such as magnetite, hornblende, garnet, pyroxenes, etc. (A. *Vema* 7-55 (355 cm); B. *Vema* 7-48 (360 cm), etc; C. *Vema* 7-44 (130 cm); D. *Vema* 18-374 (710 cm); E. *Vema* 18-374 (380 cm); F. *Vema* 16-229 (940 cm)

England are composed almost entirely of silt and sand-sized calcareous foraminifera tests. Ice-rafted sand and pebbles may constitute as much as 20%–30% of the greenish-gray continental rise sediments in the Labrador Sea and Grand Banks.

An examination of deep sea sediment cores reveal that, at least within the upper ten meters, the greatest numbers of clastic beds per given length of core

(Figure 11) occur on the continental rise (50 to 500 per 10 meters of core), whereas, the unit frequency of clastic beds is comparatively low on the continental shelf (<5 beds per 10 m), continental slope (<10 beds per 10 m) and abyssal plains (usually <50 per 10 m). Clastic beds on the continental rise (Figure 12) are generally rather thin (maximum thicknesses rarely exceed 20–30 cm and the average is < 1 cm and constitute a small percentage (usually <20%) of the total sediment column. Virtually all cores containing over 50% bedded coarse clastic material (Figure 13) and containing clastic beds over 1 meter in thickness were taken from the abyssal plains (Figure 14) (or submarine canyons (Table 1).

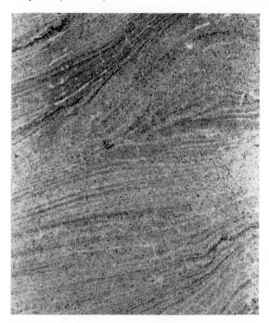

Figure 14 Laminations accentuated by concentrations of lutite in a clastic sand bed from the Sohm Abyssal Plain (× 1, photographs by direct projection of thin section). Most laminations in abyssal plain sand and silt beds are composed of alternating clay rich and clay poor layers. Concentrations of heavy minerals (placers) are infrequent. *Vema* 7-58 (428 cm)

On the continental rise, rates of accumulation of the Recent are particularly high (up to 100 cm/1000 years); with an average of about 4 cm/1000 years, and rates as high as 20 cm/1000 years predominated during the last glacial stage. Thicknesses of the Recent range between 10 and 100 cm (Figure 15) and thicknesses of the last glacial sediments range from 150 to over 1000 cm (Ericson *et al.*, 1961). As depth increases toward the abyssal plain, the rates of accumulation generally decrease; and thicknesses of the Recent drop

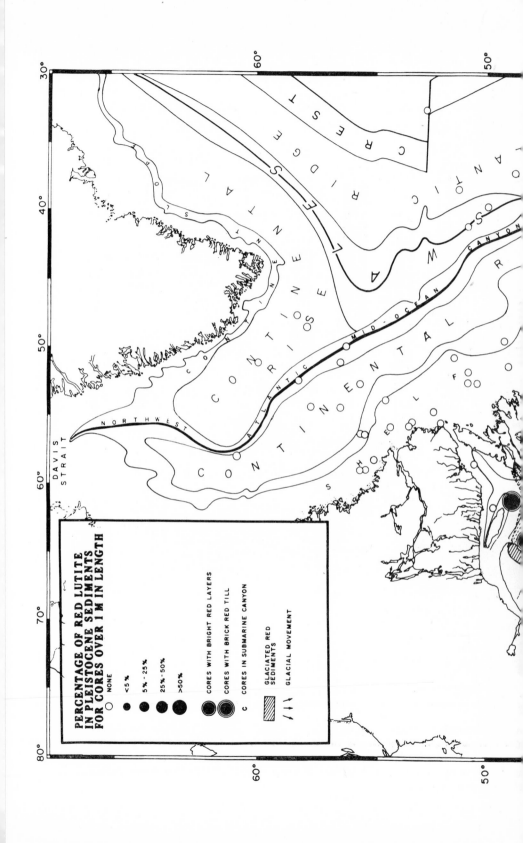

PERCENTAGE OF RED LUTITE
IN PLEISTOCENE SEDIMENTS
FOR CORES OVER 1 M IN LENGTH

○ NONE
• <5 %
● 5% - 25%
● 25% - 50%
● >50%

◉ CORES WITH BRIGHT RED LAYERS
◎ CORES WITH BRICK RED TILL
C CORES IN SUBMARINE CANYON

▨ GLACIATED RED SEDIMENTS

⚑ GLACIAL MOVEMENT

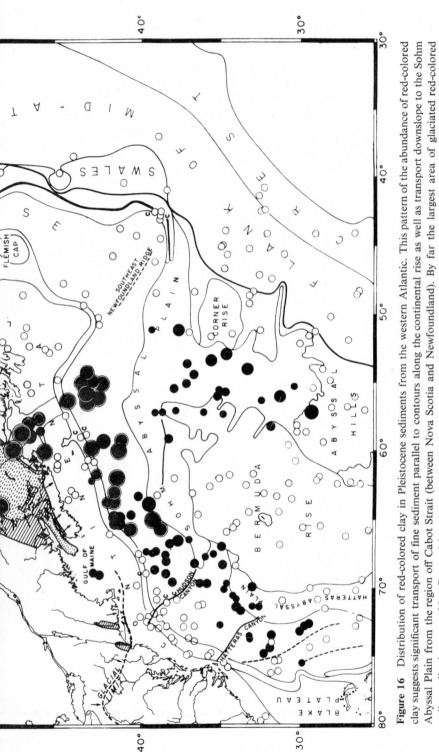

Figure 16 Distribution of red-colored clay in Pleistocene sediments from the western Atlantic. This pattern of the abundance of red-colored clay suggests significant transport of fine sediment parallel to contours along the continental rise as well as transport downslope to the Sohm Abyssal Plain from the region off Cabot Strait (between Nova Scotia and Newfoundland). By far the largest area of glaciated red-colored sediment lies in eastern New Brunswick and the Gulf of St. Lawrence and it seems likely that vast quantities of this bright-red, soft argillaceous material was carried by glaciers and outwash streams to the continental slope, dumped into the southwesterly flowing deep circulation and subsequently deposited along the continental rise as far south as 30° N by contour following currents. Turbidity currents flowing downslope from the Grand Banks off Cabot Strait (Heezen and Ewing, 1952; Heezen and Drake, 1964) probably carried some of this red-colored material onto the Sohm Abyssal Plain

from as much as 5 meters at the base of the continental slope to less than ¼ meter on the abyssal plain (Figure 8).

On the continental rise off Cabot Strait between Newfoundland and Nova Scotia the lithology of Pleistocene sediments changes abruptly (Figure 15). East of the Cabot Strait, the predominant sediment is brown, gray or very light pink silty clay, whereas, southwest of the Cabot Strait, Pleistocene sediment is predominantly brick-red clay interbedded with numerous (up to 500 per 10 meters of core) less than 1 cm thick, fine sand and silt laminations. This bright red to rose gray clay is found only in glacial age sediments.

This distinctly colored fine-grained sediment from the continental margin of the Western North Atlantic were thought to inherit their peculiar color from nearby continental sources (Ericson *et al.*, 1961). Subsequently, brick-red Pleistocene tills were reported from the Laurentian Channel (Heezen, B.C. and Drake, C.L., 1964; Conolly, J.R., Needham, H.D. and Heezen, B.C., 1967). These tills were thought to represent glacial erosion of the red Upper Paleozoic red-beds of New Brunswick and Nova Scotia. It was later suggested (D. Needham, pers. comm. 1964) that the red deep-sea sediment from the continental margin to the south came from this region.

The thickness percentage of this characteristic brick-red lutite within Pleistocene sediments on the continental rise decreases from greater than 50% off Cabot Strait to less than 10% off Cape Hatteras (Figure 16). It is also found interbedded with coarse sand beds on the Sohm Abyssal Plain as far as 1000 km south of the Gulf of St. Lawrence. The Western Boundary Undercurrent is believed to be the agent transporting this distinctive red material parallel to bathymetric contours from the continental margin off Nova Scotia to the Bahamas (Heezen, Hollister and Ruddiman, 1966) and palynologic data has further confirmed this theory (Needham, Habib and Heezen, 1969). Turbidity currents from the region of the Cabot Strait have probably transported the red sediment onto the Sohm Abyssal Plain.

TURBIDITY CURRENTS AND SLUMPS

It is now a common belief that the continental rise lies over an ancient deep-sea trench that is filled in large measure by turbidity current deposits. A turbidity current is a turbulent suspension of sediment completely mixed in water which flows downslope on the bottom of the body of water under the influence of gravity by virtue of the fact that its density is greater than that of the water above. Turbidity currents composed of high proportions of sand and silt may attain high velocities measured in km/hr and high densities exceeding 1.5 gm/cc (Kuenen and Migliorini, 1950).

Turbidity currents occur in the modern ocean and they probably transport large masses of sediment downslope from the continental slopes to the abyss. They can be considered responsible for the construction of the deep-sea cones and fans, the smoothing of the abyssal plains, the building of deep-sea channels and probably for the erosion of the deep rock-walled submarine canyons of the continental slope.

Considerable emphasis has been placed on the role of turbidity currents in the construction of the continental rise. The occurrence of turbidity currents on the continental rise is strongly suggested by the natural-leveed submarine canyons such as the Hudson and Hatteras Canyons which cut across the continental margin from shelf breaks to abyssal plain, and by the evidence of contemporary turbidity currents and their deposits such as the Grand Banks turbidity current and its extensive deposit on the abyssal plain to the south (Heezen and Ewing, 1952). Nevertheless, turbidity current deposition cannot entirely account for the dramatic contrast in physiography and sediment character between the continental rise and abyssal plains.

Seismic reflection profiles have suggested to some oceanographers (Emery *et al.*, 1970) that continental rise sediments were predominantly implaced by turbidity currents and that these deposits were further shaped by extensive contemporaneous and post depositional slumping. Gravitational sliding was given the responsibility for the shape of the characteristic convex-up accumulation of sediment that forms the upper continental rise off New York and New Jersey, as well as for the large wave like hills found at the base of the continental rise. Our alternate interpretation of the sediment bodies delineated by the seismic reflection profiles is that turbidity currents interacting with contour currents over the past 100 million years accounts for the shaping of the principal continental rise sediment bodies and that slumping on the continental rise has played a negligible role.

Continental rises constructed during successive units of geologic time have been built one upon the other off New England, each succeeding one lying seaward of the one built before (Figure 17). At the landward edge of the modern upper continental rise strata lap against the underlying beds of the continental slope whereas, on the seaward edge (at the inner margin of the lower continental rise) the same strata have been repeatedly eroded and redeposited. Lower continental rise strata lie unconformably on the truncated seaward edges of upper continental rise strata. The lack of thick sand beds or disturbed stratification in Tertiary clays recovered by the Deep Sea Drilling Project (Hollister *et al.*, 1972) from the lower continental rise indicates that turbidite sand deposition did not play a significant role in the construction of this portion of the lower continental rise.

Some workers (Ballard, 1966; Emery *et al.*, 1970) have suggested that

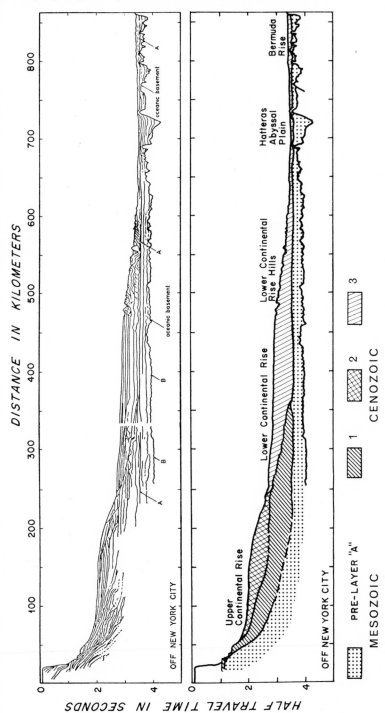

Figure 17 The continental rise off New York is composed of a seaward-migrating progression of continental rises stacked against the base of the continental slope. Turbidity currents and pelagic sources provide the necessary sediment; the Western Boundary Undercurrent subsequently and penicontemporaneously smoothes and shapes, occasionally eroding or depositing fine-grained hemipelagic continental rise sediment. Upper profile from Emery et al. (1970)

the Lower Continental Rise Hills are also produced by slumping; yet, the observation that these hills are oriented obliquely to the regional slope (Fox, Harian and Heezen, 1968), or even at right angles to the slope (Rona, 1969) suggests that they are not simple rotational slump scars. Their cross-bedded internal structure and rhythmic dune-like form suggests that they may be the product of controlled deposition by the Western Boundary Undercurrent (Fox, Harian and Heezen, 1968).

Too great an emphasis has been placed on turbidity current deposition in the explanation of sediment bodies on the continental rise. Whereas turbidity currents can be expected to be intermittent even during their maximum development, the well-known circulation of the North Atlantic Deep Water flows continuously, essentially parallel to regional contours. The finely divided sediments discharged by rivers on eastern North America, Labrador and Greenland continually reach the sea and can be swept without pause toward the southwest.

Today marine geologists have the unique opportunity of studying the products of comparatively well-known processes of deposition. In 1929, an earthquake triggered what was later identified as a turbidity current on the continental slope off Cabot Strait (Heezen and Ewing, 1952). This turbidity current deposited a graded bed of fine sand, silt and silty clay on nearly 100,000 km² of the Sohm Abyssal Plain, and this characteristic type of sediment—transported and deposited by turbidity current—is called *Turbidite*. Turbidites on the Sohm Abyssal Plain have the following characteristics: graded; moderately to poorly-sorted; grain size ranges from gravel to clay; current laminations are accentuated by concentrations of clay. *Turbidite* contains at least 10% fine ($<2 \mu$) interstitial matrix. Well preserved plant and shell fragments are common. Bottom bedding contacts are sharp and upper contacts are poorly defined.

In the preceding sections we have suggested that contour currents transport and deposit sediment on the continental rise. This sediment was found to contrast markedly with the *turbidites* (or the turbidity-current-deposited sediment) and thus, if we continue to use the term *turbidite*, it may be necessary to employ a different term for contour current-deposited sediment—*contourite*.

The deposits of contour currents, or *contourite*, on the continental rise of North America have the following characteristics: grading either normal or reversed; well to very well sorted; grain size is predominantly fine sand, silt or lutite; beds are usually 0.1 to 10 cm thick; upper and lower contacts are sharp; abundant laminations are accentuated by concentrations of heavy minerals; preferred grain orientation is parallel to the bedding plane; interstitial matrix is less than 5%. Microfossils tend to be worn or broken.

Table 2 Principal Characteristics of *Turbidite* and *Contourite* Sand or Silt

		Turbidite	Contourite	Conclusions
Size Sorting		moderate to poorly sorted > 1.50 (Folk)	well to very well sorted < 0.75 (Folk)	contourite is better sorted
Bed Thickness		usually 10–100 cm	usually < 5 cm	contourite has thinner bedding
Primary sedimentary structure	Grading	normal grading ubiquitous, bottom contacts sharp, upper contacts poorly defined	normal and reverse grading, bottom and top contacts sharp .	contourite tends to be less regularly graded and has sharp upper contacts
	Cross laminations	common, accentuated by concentrations of lutite	common, accentuated by concentration of heavy minerals	contourite contrasts sharply with turbidite in that heavy mineral placers are in the form of small scale stratification
	Horizontal laminations	common in upper portion only, accentuated by concentration of lutite	common throughout, accentuated by concentrations of heavy minerals or foraminifera shells	
Massive Bedding		common, particularly in lower portion	absent	contourite is ubiquitously laminated
Grain Fabric		little or no preferred grain orientation in massive graded portions	preferred grain orientation parallel to the bedding plane is ubiquitous throughout bed	contourite has better grain orientation
Principal constituents of sand and silt beds	Matrix ($< 2\mu$)	10 to 20%	0 to 5%	contourite has less matrix
	Microfossils	common and well preserved, sorted by size throughout bed	rare and usually worn or broken, often size sorted in placers	contourite shows more evidence of reworking
	Plant and skeletal remains	common and well preserved, sorted by size throughout bed	rare and usually worn or broken	contourite shows more evidence of reworking
Classification (Pettijohn)		graywacke and sub-graywacke	sub-graywacke, Arkose and orthoquartzite	contourite is more "mature"

The use of such generic terms as *contourite* or *turbidite* (Table 2) in descriptions of ancient deposits seems ill-advised, however. Kuenen (1964) "insists"—"that no single feature is diagnostic of turbidites, for each can be encountered in deposits of quite different nature," and grants that "it is seldom possible to be entirely sure whether a (particular) bed is a turbidite or not". Although he concludes that "a large number, probably the great majority, of flysch formations are built up entirely or for the major part by turbidite sequences", he states that "modern deep sea sands differ from most of the typical flysch rocks in more than one way".

His view that "convincing evidence for turbidites is only obtained if a sufficient number of positive structures is combined with the absence of shallow water features," implies that the variety of primary structures attributed to turbidity current deposition cannot be formed by any other process in the deep sea. Since we have established the sedimentological importance of deep contour currents in great ocean depths it is abundantly clear that the demonstration of a deep water environment is alone insufficient evidence to establish a particular clastic bed as the product of turbidity current deposition.

Many forms of current structures have been reported from bedding planes of ancient flysch sequences. (See reviews in Bouma and Brouwer, 1964; Bouma, 1962; Dzulynski and Walton, 1965; Potter and Pettijohn, 1963; Pettijohn and Potter, 1964). These features are generally observed as casts on the underside of sandstone beds. Bedding plane current structures, other than ripple marks, are generally classified as flutes, grooves, crescents, shadows, bounce, brush, prod and slide marks. In most cases where the features are well expressed the direction of current is clearly indicated. Current structures formed by depositing turbidity currents cannot be observed on the modern deep-sea floor, since they are always covered by the "turbidite" deposit. However, current structures such as sand shadows (Potter and Pettijohn, 1963) and current crescents or scour-remnant ridges (Allen, 1965) constructed by traction transport and deposition from deep-sea currents are frequently observed on the deep-sea floor and since these positive constructional features can be load casted into the underlying clay, they may also appear as casts in ancient rocks.

Many of the above-mentioned current structures have been attributed to turbidity currents, and it has often been assumed that they are almost diagnostic of "turbidites". Although we cannot at present determine which, if any, of these primary bedding plane structures are created by modern turbidity currents, deep-sea photographs conclusively demonstrate that many if not all are being constructed at present by contour currents. Bouma (1962) inferred that the ideal "turbidite" would, if complete, contain an invariable sequence of primary structures; these are, from bottom to top: 1) lowermost graded interval "a"; 2) horizontally laminated interval "b"; 3) convolute and cross laminated interval 'c"; 4) horizontally laminated interval "d"; and 5) pelagic lutite interval "e". However, in his study of the Peira-Cava area in the Maritime Alps of southeastern France he rarely found the complete sequence (a–e). A sequence sometimes observed in deep-sea sands and silts consists of a graded and often laminated interval overlain by a cross-laminated interval beneath pelagic lutite (i.e., "a" + "b", "c", "e").

Bouma postulated that with increase in distance from the source the sequence shortens as intervals "a" through "d" disappear sequentially. However, the absence of the upper part of the sequence may be explained by "erosion by a subsequent current". Thus the frequent occurrence of the lower intervals of the sequence a, b, e. ... would suggest a proximal source whereas the absence of the lower intervals would indicate a distant source. This leads to the contradiction, if applied to sediments of the modern deep-sea floor, that the abyssal plains are closer to the source than the continental rise.

The present study has demonstrated that many if not all of the primary structures generally listed as characteristic of ancient "turbidites" are also characteristic features of modern deposits of contour currents or contourites, and the application to the modern sea floor of Bouma's hypothesis of sequential basal truncation with increasing distance from the source leads to impossible conclusions.

CONCLUSIONS

Near-bottom velocities of 10 to 25 cm/sec and higher have been observed beneath the Western Boundary Undercurrent from Greenland to Cape Hatteras. These measured velocities are believed to be competent to transport many of the sediment sizes found on the continental rise and thus it is argued here that transportation and deposition of sediment along the continental slope and continental rise of the Western North Atlantic should be strongly affected by this contour-following deep-sea current.

It is necessary, however, to appeal to some process that will supply sediment to this dynamic system from the edge of the continents. Submarine canyons are most likely a product of erosion and transportation of detritus by turbidity currents (Daly 1936). These same currents provide a mechanism for the injection of sediment into contour-following bottom currents. It seems likely that the finer sediment carried downslope by turbidity currents is picked up in the contour currents. Silt carried in suspension by the turbidity current would be deposited on the gentler portions of the continental rise and only the very largest and densest turbidity currents could continue across the continental rise and deposit their coarser load on the flat abyssal plains.

Deep contour currents are deflected against the side of the oceanic basin. The lowest level at which such a current can effectively flow lies near the margin of the nearby level abyssal plains. The combination of turbidity currents and contour currents can account for:

1   The flat-lying beds of coarse detritus on the abyssal plains. Only the largest and most competent turbidity currents can flow across the continental rise and deposit their load on the flat abyssal plains.

2   Current lineations and ripples on the continental rise show that currents flow parallel to, not normal to, bathymetric contours; and that flow is in the same general direction as the over-lying deep and bottom water.

3   Relatively poor echo reflectivity of the continental rise, and the relatively good echo reflectivity of the abyssal plains.

4   The continental rise with numerous (as many as 500 per 10 meters of core) relatively clean (less than 10% interstitial matrix), heavy mineral placered, cross-bedded silt and fine sand beds (0.1 to 10 cm thick) with well-developed grain fabric and the abyssal plain sediment which often contains thick (>50 cm) muddy (often as high as 40% interstitial matrix), massive and horizontally laminated sand beds with little or no grain fabric.

5   Transport of continental rise sediment parallel to the bathymetric contours for at least 1500 km as demonstrated by the construction of the Blake-Bahama Outer Ridge. In addition, the distribution of brick-red-colored clay in glacial-age sediment in the western North Atlantic indicates massive sediment transport parallel to bathymetric contours along the continental rise from the Cabot Strait to the Blake-Bahama Outer Ridge (over 3000 km) and downslope from the Cabot Strait to the southern end of the Sohm Abyssal Plain (over 1000 km).

References

Ballard, J. A., Structure of the lower continental rise hills of the western North Atlantic. *Geophysics*, **31**, 506–523, 1966.

Bouma, A. H., *Sedimentology of Some Flysch Deposits*. Elsevier Publishing Co., Amsterdam, 148 pp., 1962.

Bouma, A. H. and A. Brouwer (Editors), *Turbidites*. Elsevier Publishing Company, Amsterdam, 264 pp., 1964.

Conolly, J. F., H. D. Needham, and B. C. Heezen, Late Pleistocene and Holocene sedimentation in the Laurentian Channel. *J. Geol.*, **7**, 131–147, 1967.

Daly, R. A., Origin of submarine canyons. *Amer. J. Sci.*, **31**, 401–420, 1936.

Dzulnski, S. and E. K. Walton (Eds.), *Sedimentary Features of Flysch and Greywackes*. Elsevier Publishing Co., Amsterdam, 265 pp., 1965.

Eittrem, S., M.Ewing, and E.M.Thorndike, Suspended matter along the continental margin of the North American Basin. *Deep-Sea Res.*, **16**, 613–624, 1969.

Elmendorf, C.H. and B.C.Heezen, Oceanographic information for engineering submarine cable systems. *The Bell System Technical Journal*, XXXVI, 1047–1093, 1957.

Emery, K.O. and D.M.Owen, Current markings on the continental slope. In: Hersey, J.B. (Editor), Deep Sea Photography. *The Johns Hopkins Oceanographic Studies*, **3**, 167–172, 1967.

Emery, K.O., E.Uchupi, J.D.Phillips, C.O.Bowin, E.T.Bunce, and S.T.Knott, Continental rise off eastern North America. *Amer. Assoc. Petr. Geol. Bull.*, **54**, 44–108, 1970.

Ericson, D.B., M.Ewing, G.Wollin, and B.C.Heezen, Atlantic deep-sea sediment cores. *Bull. Geol. Soc. Amer.*, **72**, 193–286, 1961.

Ewing, M. and F.Mouzo, Ocean bottom photographs in the area of the oldest known outcrops, North Atlantic Ocean. *Proc. Nat'l. Acad. Sci.*, **61**, 787–793, 1968.

Fox, P.J., A.Harian, and B.C.Heezen, Abyssal Anti-dunes. *Nature*, **220**, 470–472, 1968.

Heezen, B.C. and C.L.Drake, Grand Banks slump. *Am. Assoc. Petrol. Geol.*, **48**, 221–233, 1964.

Heezen, B.C. and M.Ewing, Turbidity currents and submarine slumps and the 1929 Grand Banks earthquake. *Am. J. Sci.*, **250**, 849–873, 1952.

Heezen, B.C. and C.D.Hollister, Evidence of deep-sea bottom currents from abyssal sediments (abs.). *Abstracts of papers, Internat. Assoc. of Phys. Oceanog., 13 th General Assembly, Internat. Union Geodesy and Geophy.*, **6**, 111, 1963.

Heezen, B.C. and C.D.Hollister, Deep-sea current evidence from abyssal sediments. *Marine Geology*, **1**, 141–174, 1964.

Heezen, B.C., C.D.Hollister, and W.F.Ruddiman, Shaping of the continental rise by deep geostrophic contour currents. *Science*, **152**, 502–508, 1966.

Heezen, B.C. and C.D.Hollister, *The Face of the Deep*. Oxford University Press, New York, 650 pp., 1971.

Heezen, B.C. and G.L.Johnson, Mediterranean Undercurrent and Microphysiography west of Gibraltar. *Bull. Inst. Oceanog., Monaco*, **67** (1382), 95 pp., 1969.

Heezen, B.C., E.D.Schneider, and O.H.Pilkey, Sediment transport by the Antarctic Bottom Current on Bermuda Rise. *Nature*, **211**, 611–612, 1966.

Hollister, C.D., Sediment distribution and deep circulation in the western North Atlantic. Unpublished *Ph. D. Dissertation*, Columbia University, New York, 1967.

Hollister, C.D. and R.B.Elder, Contour currents in the Weddell Sea. *Deep-Sea Res.*, **16**, 99–101, 1969.

Hollister, C.D. and B.C.Heezen, The Floor of the Bellingshausen Sea. *Johns Hopkins Oceanographic Studies*, **3**, 177–189, 1967.

Hollister, C.D. *et al.*, *Initial Reports of the Deep Sea Drilling Project*, **XI**, U.S. Government Printing Office, Wash. D.C. 1972.

Hjulstrom, F., Studies of the morphological activities of rivers illustrated by the River Fyris. *Bull. Geol. Inst. Univ. Upsala*, **25**, 221–527, 1935.

Jones, E.J.W., M.Ewing, J.I.Ewing, and S.L.Eittrem, Influences of Norwegian Sea overflow water on sedimentation in the northern North Atlantic and Labrador Sea. *Jour. Geophys. Res.*, **75**, 1655–1680, 1970.

Knott, S.T. and H.Hoskins, Evidence of Pleistocene events in the structure of the continental shelf off the northeastern United States. *Marine Geology*, **6**, 5–43, 1968.

Kuenen, Ph.H., Deep-sea sands and ancient turbidites. In: Bouma, A.H. and A.Brouwer (Eds.), *Turbidites*, Elsevier, Amsterdam, pp. 3–33, 1964.

Kuenen, Ph. H., Experiments in connection with turbidity currents and clay-suspensions. In: Whittard, W. F. and R. Bradshaw (Eds.), *Submarine Geol. and Geophys., Colston Papers*, **17**, 47–74, 1965.

Kuenen, Ph. H. and C. I. Migliorini, Turbidity currents as a cause of graded bedding. *J. Geol.*, **58**, 1–127, 1950.

Laughton, A. S., Discrete hyperbolic echoes from an otherwise smooth deep-sea floor. *Deep-Sea Res.*, **9**, 218–219, 1962.

Mavis, F. T., C. Ho, and Y. Tu, The Transportation of Detritus by Flowing Water I. *Studies in Engineering*, Univ. Press, Iowa City, Iowa, 1935.

McCoy, F. W., Jr., Bottom currents in the western Atlantic Ocean between the Lesser Antilles and the Mid-Atlantic Ridge. *Deep-Sea Res.*, **15**, 179–184, 1968.

Needham, H. D., D. Habib, and B. C. Heezen, Upper Carboniferous palynomorphs as a tracer of red sediment dispersal patterns in the Northwest Atlantic. *Journal of Geology*, **77**, 113–120, 1969.

Pettijohn, F. P. and P. E. Potter, *Atlas and glossary of primary sedimentary structures*. Springer-Verlag, New York, 370 pp., 1964.

Potter, P. E. and F. P. Pettijohn, *Paleocurrents and basin analysis*. Academic Press, New York, 295 pp., 1963.

Rees, A. I., Some flume experiments with fine silt. *Sedimentology*, **6**, 209–240, 1966.

Rona, P. A., Linear "Lower Continental Rise Hills" off Cape Hatteras. *Jour. Sed. Pet.*, **39**, 1132–1141, 1969.

Schneider, E. D., P. J. Fox, C. D. Hollister, D. Needham, and B. C. Heezen, Further evidence for contour currents in the western North Atlantic. *Earth and Planetary Sciences Letters*, **2**, 351–357, 1967.

Southard, J. B., Young, R. A., Hollister, C. D., Experimental Erosion of Calcareous Ooze. *Jour. Geophys. Res.*, **76**, 5903–5910, 1971.

Swallow, J. C. and L. V. Worthington, An observation of a deep countercurrent in the western North Atlantic. *Deep-Sea Res.*, **8**, 1–19, 1961.

Swallow, J. C. and L. V. Worthington, Deep currents in the Labrador Sea. *Deep-Sea Res.*, **9**, 493–500, 1969.

Volkmann, G., Deep current observations in the western North Atlantic. *Deep-Sea Res.*, **9**, 493–500, 1962.

Wüst, G., Schichtung und Zirkulation des Atlantischen Ozeans. Das Bodenwasser und die Stratosphäre. *Wiss. Erg. Deutsch. Atlant. "Meteor" 1925–1927*, **6**, 1–288, 1936.

Wüst, G., Die Stromgeschwindigkeiten und Strommengen in der Atlantischen Tiefsee. *Geol. Rundschau*, **47**, 187–195, 1958.

Oceanographic Measurements from Anchored and Dynamically-Positioned Ships in Deep Water*

ROBERT D. GERARD

Lamont-Doherty Geological Observatory, Columbia University
Palisades, New York 10964

Abstract   An account is given of the current measurements obtained by anchored ships at deep-ocean stations since Pillsbury's initial work in 1885. Oceanographic measurements resulting from the operations of dynamically-positioned drilling ships in the deep sea are described, including new subsurface current measurements in the western Atlantic. The station-keeping capabilities of these ships are described, and suggestions are given for enlarging their role in taking modern oceanographic observations.

The goal of obtaining direct measurement of ocean currents from a ship anchored in the deep ocean was first successfully accomplished in the decade following the *Challenger* expedition. On a series of expeditions between 1885 and 1890 the U.S. Coast and Geodetic Survey Vessel *Blake*, under the command of Lt. J.E.Pillsbury, U.S.N. (except for the year 1890, when the *Blake* was commanded by Lt. C.E.Vreeland, U.S.N.), carried out thousands of current measurements (from surface to a maximum of 400 m) to determine the nature of the Gulf Stream and its origins. Pillsbury accomplished these difficult measurements using current meters and anchoring equipment of his own design.

This work and that of subsequent expeditions on which deep anchor stations were made have been summarized by Bowden (1954), but his Table 1 (page 36) requires certain additions. With regard to the work of Pillsbury, Bowden lists 39 stations with duration of observations "up to

---

* Lamont-Doherty Geological Observatory, Contribution Number 1621.

166 hours". The reports of the U.S. Coast and Geodetic Survey between 1886 and 1891 (Pillsbury, 1886, 1887, 1888, 1890, 1891) show that at least 164 anchorages (the deepest being 3987 m) were made by the *Blake* at some 90 stations deeper than 350 m between Cape Hatteras and the southernmost passages of the Lesser Antilles. One station in the Florida Current was occupied 22 times during several years for a total of more than 600 hours of current observations.

The Research Vessel *Meteor* is credited with 12 anchor stations (9 during the German Atlantic Expedition of 1925–27 and 3 during the German North Atlantic Expeditions of 1937 and 1938). To this total should be added current measurements from 7 anchor stations of the 1937 expedition which were lost during World War II and have recently come to light (Tomczak, 1970).

Bowden mentions, but does not list, the number of stations made in the Pacific on the *Snellius* Expedition of 1929–30. There were, in fact, 9 stations resulting from this work (Lek, 1938). Thus, in the work of all oceanographic expeditions from 1895 to 1938, some 211 deep-ocean anchored current-meter stations should be listed. It is astonishing that Pillsbury's work in the 19th century represents 70 per cent of this total.

It should be mentioned that the aims of the later workers in making observations of current in the open ocean were different from those of Pillsbury, who was concerned with the measurement of strong, permanent currents in the Gulf Stream system. The open ocean measurements were made in an attempt to interpret the relative field of currents and confirm the dynamic methods of Sandström and Helland-Hansen (1903) for computing the geostrophic current.

Due to uncertainties caused by ship's motion (overriding, swing, and yaw) and the generally small current speeds encountered, anchor station measurements have been mainly useful in interpreting tides in the deep ocean (Defant, 1932; Helland-Hansen, 1930; Thorade, 1934; Schubert, 1944).

In one notable study, however, Wüst (1924) was able to compute an absolute current profile across the Florida Straits based upon Pillsbury's direct determination of a reference level. The correspondence which he found between currents computed from the observed field of mass and the observed current field was a convincing demonstration of the dynamic methods.

Improvements both in theory and technology have lessened the need for deep-sea anchored-station current observations in recent years. The application of Defant's (1941) method for indirectly determining, rather than observing, a reference level (assuming that a level of "no horizontal motion" coincides with a subsurface zone where constant relative pressure differences exist between adjacent oceanographic stations) has led to reasonable interpretations of geostrophic current patterns over broad ocean areas (Wüst, 1957; Gordon, 1967).

The use of modern marine radar to measure the movement of a drifting ship from fixed reference buoys has allowed direct measurement of well-defined near-surface ocean currents in the open sea using deck-lowered current meters (Knauss, 1960, 1966). Similar methods have also been successful in the deep ocean to measure subsurface currents with drogues (Gerard *et al.*, 1965).

Long-term Eulerian measurements of deep currents are today routinely obtained using anchored current-meter arrays developed by Richardson (1963). Lagrangian measurements of deep currents, using acoustical methods for tracking the trajectories of neutrally-buoyant floats developed by Swallow (1955) are also commonly made.

Deep anchorages are still occasionally made by modern oceanographic vessels to obtain current measurements. Gerard (1963, 1964) reports twelve anchored current-meter stations in 2000-m depths in the passages of the Caribbean-Antillean area. New sampling requirements and methods of measurement of ocean parameters other than currents have placed renewed emphasis on ships which can remain stationary at deep ocean positions. Beck and Ess (1962) have described deep anchorages in the North Atlantic for the purpose of acoustical measurements. Five ships moored to anchored buoys in 6000-m depths were employed in the Barbados Oceanographic and Meteorological Expedition (BOMEX) in 1969 to obtain time-series oceanographic and meteorological data in the western Tropical Atlantic.

The exploitation of offshore oil resources in recent years has given rise to a whole new technology for positioning ships at fixed locations in deep water. The first application of these techniques in water depths greater than 1000 m took place in the Pacific Ocean off the coast of California during the preliminary phase of the Mohole Project in 1961 (Horton, 1961; AMSOC Committee of the Division of Earth Sciences, 1961). Using radar and sonar ranging on an array of taut-line buoys (Figure 1), the 3000-ton drilling barge *Cuss I* was kept on position dynamically within a 100-m radius, using four 200-HP motors under the control of a pilot, in water depths as great as 3500 m. Internally-recording current meters were deployed from *Cuss I* and measurements obtained down to depths of 3000 m. These measurements showed considerably greater variability than others obtained by meters suspended from nearby taut-line buoys, the difference presumably being due to maneuvering of the ship during the drilling operation. Parachute drogue measurements of surface currents were also made from the dynamically-positioned ship with good results.

The next operation of a dynamically-positioned ship which afforded op-

portunity for current measurements was the Blake Plateau drilling project
under the aegis of the Joint Oceanographic Institutions Deep Earth Sampling (JOIDES) program, on which the author participated. During April
and May 1965, the Motor Vessel *Caldrill I*, equipped with rotary drilling
equipment, drilled and cored at six sites across the continental shelf and

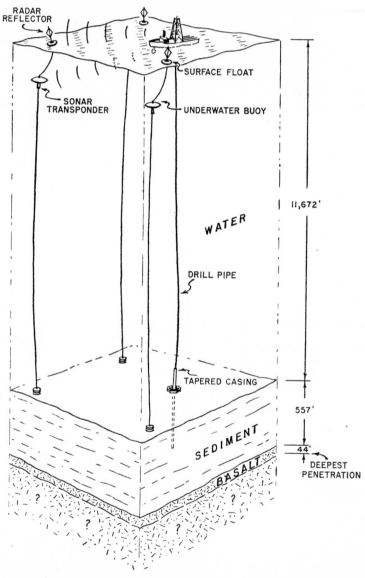

Figure 1 Schematic drawing of the drilling barge *Cuss I* at the Mohole Guadalupe
site (from AMSOC Committee of the Division of Earth Sciences, 1961)

Blake Plateau off the northern coast of Florida (Joides, 1965; Schlee and Gerard, 1965) in maximum water depths of 1032 m. The automatic position-keeping method utilized an analog computer to control four 300-HP motors. The computer, in turn, received signals from a gyro compass and an angle-sensing transducer mounted above a constant-tension taut wire anchored to the ocean bottom. In this system, if the ship drifts away from a point directly above the anchor, the transducer senses the departure of the wire from the vertical and generates signals through the computer to move the ship back to position. In this way the ship stayed on station automatically

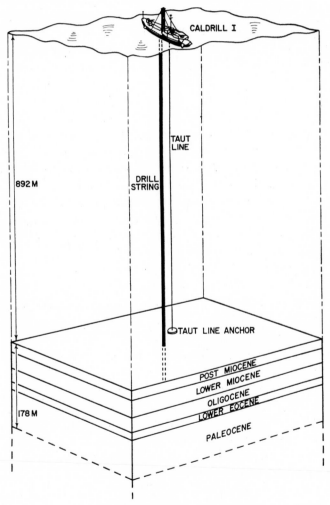

Figure 2 Schematic drawing of the drilling vessel *Caldrill I* at the Joides, Blake Plateau Project, Site 4

within a radius amounting to less than 3 per cent of the water depth. Figure 2 is a schematic of the *Caldrill I* station-keeping and drilling operation.

During the Blake Plateau project it was necessary to measure surface currents while maintaining station and drilling close to the axis of the Gulf Stream. The use of deck-lowered current meters in the near-surface waters from a dynamically-positioned ship presents special problems. Not only is the magnetic direction apparatus of the current meter strongly affected down to several tens of meters below a steel hull (Helland-Hansen, 1930), but the turbulence created by the positioning motors will strongly affect current speed measurements. Since no support vessel was available to assist in taking surface current measurements, a variation of the "chip log" method was used. With the ship maintaining station, headed into the direction of the current, a man was sent to the bow with several oranges. Upon signal from the timekeeper, he would throw an orange about 60 m perpendicular to the ship's axis. The drift of the orange parallel to the ship's heading would be timed by stopwatch for a distance nearly equal to the ship's length (50 m). Repeated measurements taken in this manner agreed within a few per cent. The use of oranges for chip-log measurements have the following advantages: near-neutral buoyancy, low windage, optimum size and mass for throwing, and high visibility. One disadvantage (observed on a more recent cruise aboard the Drilling Ship *Glomar Challenger*) is that they are edible and are occasionally eaten by sharks before current measurements can be completed. To test how well this chip-log followed the water flow, pellets of fluorescein dye were inserted in slots under the skin of the orange. Upon contact with the sea, a patch of yellow-green colored water would appear around the orange. The orange invariably remained centered in the dye patch during its trajectory past the ship.

In addition to surface current measurements, one current-meter profile (to 100-m depth), two shallow (30-m) temperature/salinity profiles, two bottom current measurements, three subsurface time-series temperature records, four bottom grab samples, and two bottom camera stations were obtained using equipment lowered from a whinch while the *Caldrill I* was on station during drilling operations. Three parachute drogue measurements of current (surface, 366-m, 380-m) were also accomplished. The average velocities observed in the two subsurface measurements are shown together with surface currents in Figure 3, and the covering data are listed in Table 1.

The success of the Blake Plateau drilling program led to a more extensive deep-sea effort called the Deep Sea Drilling Project, under the guidance of the Joides organization (Joides, 1967). In order to provide capabilities for drilling and coring in ocean depths of 6000 m, the 10,000-ton Drilling Ship *Glomar Challenger* was constructed and launched along with the project in

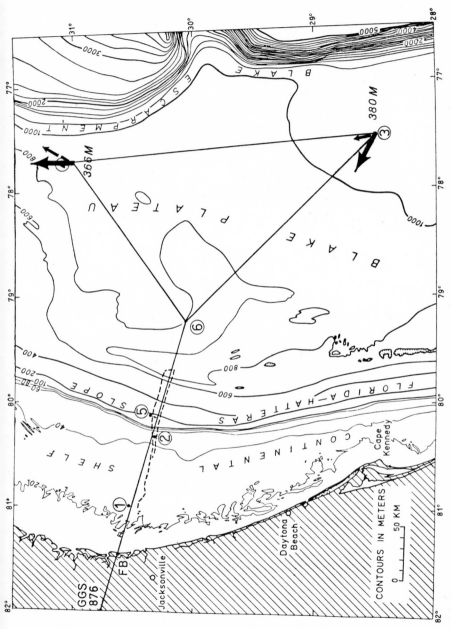

Figure 3 Blake Plateau current measurements from the M/V *Caldrill I*. Heavy arrows show surface current direction; smaller arrows show subsurface current direction by parachute drogue at the indicated depth. Lengths of arrows show relative current speed. Drilling sites (Schlee and Gerard, 1965) are identified by circled numbers

the summer of 1968. Since that time drilling and coring operations (in a maximum water depth of 5500 m) have been carried out at more than 150 sites in the Atlantic and Pacific Oceans, Gulf of Mexico, and Mediterranean and Caribbean Seas.

Table 1   Blake Plateau Parachute Drogue Current Measurements

Date 1965	Location	Ocean Depth (m)	Drogue Depth (m)	Duration of Observ. (hrs.)	Direction	Speed (cm/sec)
3–4 May	28° 30′ N 77° 31′ W	1032	366	24	352° (300°)	17 (32)
14 May	31° 03′ N 77° 45′ W	885	380	5	031° (000°)	14 (33)

(   ) indicate surface current data.

The dynamic-positioning system on the *Glomar Challenger* is described by Peterson *et al.* (1970). Four tunnel thrusters (two in the bow and two aft) are operated in conjunction with the ship's main screws and enable her to move in any direction. While on station, four hydrophones fixed beneath the hull continually receive signals from an acoustical beacon emplanted on the sea floor. A computer calculates the ship's position above the beacon from the delay times of the arriving signals and automatically controls the propulsion machinery to maintain the ship's heading and location over the drill hole. Continuous recordings of position show that under normal conditions it is possible to maintain the ship within 24 m of the desired location for periods as long as six days. A gyroscopically-controlled roll-stabilizing system limits the ship's roll to less than 5° on station.

Because of the intensive concentration on drilling and coring, the unique capabilities of the *Glomar Challenger* have thus far been little used for other oceanographic measurements. A few (mostly indirect) observations made by the author in the western Atlantic and Caribbean on Leg 4 (February–March 1969) of the Deep Sea Drilling Project suggest the potential capabilities of such ships for taking oceanographic measurements:

(1) Chip-log measurements (in the manner described above) were made at six of the nine sites during Leg 4.

(2) At Site 24 in the northern Brazil Basin, while drilling at a location "offset" by 2500 ft from the beacon, relatively large discrepancies were noted between the ship's acoustical position and that indicated by satellite navi-

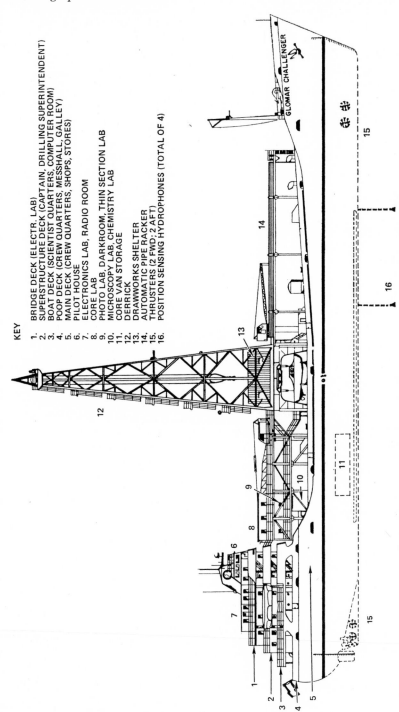

KEY

1. BRIDGE DECK (ELECTR. LAB)
2. SUPERSTRUCTURE DECK (CAPTAIN, DRILLING SUPERINTENDENT)
3. BOAT DECK (SCIENTIST QUARTERS, COMPUTER ROOM)
4. POOP DECK (CREW QUARTERS, MESSHALL, GALLEY)
5. MAIN DECK (CREW QUARTERS, SHOPS, STORES)
6. PILOT HOUSE
7. ELECTRONICS LAB, RADIO ROOM
8. CORE LAB
9. PHOTO LAB, DARKROOM, THIN SECTION LAB
10. MICROSCOPY LAB, CHEMISTRY LAB
11. CORE VAN STORAGE
12. DERRICK
13. DRAWWORKS SHELTER
14. AUTOMATIC PIPE RACKER
15. THRUSTERS (2 FWD; 2 AFT)
16. POSITION SENSING HYDROPHONES (TOTAL OF 4)

Figure 4 Profile view of the drilling ship *Glomar Challenger* (from Peterson et al., 1970)

gation fixes and drill floor indications. It is suggested that refraction of the sound path due to long-period internal waves or internal tides (La Fond, 1966), affecting the upper thermal structure, may have caused this problem. The use of a vertical thermistor array or frequent bathythermograph (or expendable bathythermograph) observations while on station would provide valuable information on time-related thermal structure fluctuations.

(3) Site 30, located at 12° 52.9′ N, 63° 23′ W on the Aves Ridge (Bader *et a.l*, 1970), lies close to the axis of the strongest surface current entering the southern Caribbean from the Atlantic (Wüst, 1964). The site is located on the east flank of an elevated portion (minimum depth about 600 m) of the north–south trending Aves Ridge, whose general crest height is about 1250 m. The ship approached this site from the west, steering and making good a course of 117°. In the vicinity of the western flank of the ridge, satellite navigation fixes showed that the ship was set strongly to the south. After crossing the shallowest part of the Aves Ridge, and while maneuvering on the east flank for the drilling station, a strong current to the northwest was observed. After arriving on station, the surface current speed proved to be 2.5 kts (129 cm/sec). The deflection of the streamlines over the ridge appears to correspond to Sverdrup's (1941) model for a deep-reaching current. In fact, the presence of a strong current at the bottom (1218-m deep) was partly responsible for prematurely terminating the drilling work at this station.

The bottom beacon is constructed with a buoyant acoustic transmitter that floats about 10-m above the ocean floor. At this station the bottom current appeared to deflect the buoyant transmitter so that the center of its directional beam (with a half-angle beam width of about 15°) was focused to the west instead of directly upward toward the ship. This caused very poor and intermittent reception of the signal and was the main cause for leaving this station after only 24 hours of drilling. It is interesting to note that Gordon *et al.* (1966) observed strong temperature fluctuations in the bottom water on the Aves Ridge (at a position 20 miles north of our station), believed due to the passage of a tropical storm. Our experience suggests the desirability of using deck-lowered current meters from the dynamically-positioned ship in strong currents and the use of bottom current sensors whose information can be telemetered acoustically along with the bottom beacon signal.

(4) At Station 31 in the central Caribbean (Gerard, 1970; Gealy and Gerard, 1970) an observation of electrical potential was made at the base of the upper mixed layer which suggests the presence of a thin (10-m) zone of oxygen-depleted water in the pycnocline in a depth of 100 m. This measurement points out the potential value of using continuously-recording probes

from a stationary ship for detailed study of fine-scale features in the stratification.

Cooper and Stommel (1968) have demonstrated the existence of remarkably regular step-like features of salinity and temperature in the main thermocline in the Sargasso Sea, based on repeated lowerings of an STD instrument over several hours from a drifting ship. Longer time-series measurements from stationary vessels would greatly improve our knowledge of the nature and origin of these apparently widespread features.

The results of the *Glomar Challenger* work leave no doubt that dynamic positioning of ships in the open sea has become a standard practice. It would be a relatively minor development to expand the existing system for taking routine current measurements throughout the entire ocean column using free-fall devices. An acoustic transmitter (using a frequency separate from that of the positioning beacon), adjusted for slightly negative buoyancy, could be deployed from the ship and the horizontal components of its motion accurately recorded simultaneously with the ship's motion.

These examples of the capabilities and potentials of dynamically-positioned ships for taking oceanographic measurements suggest that oceanographer's ideal of a fixed observation platform in the deep ocean may soon be realized.

Acknowledgments

The *Caldrill I* and *Glomar Challenger* drilling operations reported in this paper were funded by National Science Foundation Grant GP-4233 and Contract C-482, respectively. The current measurements and other oceanographic observations of the Blake Plateau Project were supported by the U.S. Atomic Energy Commission Contract AT(30-1)2663.

References

AMSOC Committee of the Division of Earth Sciences, National Research Council, *Experimental Drilling in Deep Water at LaJolla and Guadalupe Sites*, National Academy of Sciences/National Research Council Publication No.914, 183 pp., Washington, D.C., 1961.

Bader, R.G., R.D.Gerard, W.E.Benson, H.M.Bolli, W.W.Hay, W.T.Rothwell, Jr., M.H.Ruef, W.R.Riedel, and F.L.Sayles, *Initial Reports of the Deep Sea Drilling Project*, Vol.IV, 753 pp., Washington (U.S. Government Printing Office), 1970.

Beck, H.C. and J.O.Ess, Deep-sea anchoring, Hudson Laboratories of Columbia University Tech. Report No.98, 64 pp., 1962.

Bowden, K.F., The direct measurement of subsurface currents in the oceans, *Deep-Sea Res.*, **2**, 33–47, 1954.

Cooper, J.W. and H.Stommel, Regularly spaced steps in the main thermocline near Bermuda, *J. Geophys. Res.*, **73**, 5849–5854, 1968.

Defant, A., Die Gezeiten und inneren Gezeitenwellen des Atlantischen Ozeans, "*Meteor*" *Reports*, Bd.VII, 1. Teil, 318 pp., 1932.

Defant, A., Die absolute Topographie des physikalischen Meeresniveaus und die Druck-flächen sowie die Wasserbewegung im Atlantischen Ozean, *"Meteor" Reports*, Bd. VI, 2. Teil, 5 Lief., 191–260, 1941.

Gealy, E. and R. Gerard, In situ petrophysical measurements in the Caribbean, in *Initial Reports of the Deep Sea Drilling Project*, Vol. IV, 267–293, U.S. Government Printing Office, Washington, D.C., 1970.

Gerard, R., Direct current measurements and hydrographic observations in the Caribbean-Antillean region, Lamont Geological Observatory Tech. Report CU-5-63, 21 pp. (unpublished), 1963.

Gerard, R., Caribbean current measurements, July 1964, Lamont Geological Observatory Tech. Report CU-2663-14, 17 pp., (unpublished), 1964.

Gerard, R., R. K. Sexton, and P. Majeika, Parachute drogue measurements in the Eastern Tropical Atlantic in September 1964, *J. Geophys. Res.*, **70**, 5696–5698, 1965.

Gerard, R., A problematical measurement of electrical potential in the upper 300 meters of the central Caribbean, *Marine Technology* 1970, *Preprints*, Vol. 2, 1433–1444, Marine Technology Society, Washington, D.C., 1970.

Gordon, A. L., P. J. Grim, and M. Langseth, Layer of abnormally cold bottom water over southern Aves Ridge, *Science*, **151**, 1525–1526, 1966.

Gordon, A. L., Circulation of the Caribbean Sea, *J. Geophys. Res.*, **72**, 6207–6223, 1967.

Helland-Hansen, B., Physical oceanography and meteorology, *"Michael Sars" Exped.* 1910, *Report*, Vol. I, Pt. I, 115 pp., Pt. 2, 102 pp., 1930.

Horton, E. E., Preliminary drilling phase of Mohole Project, I. Summary of drilling operations, *Bull. Am. Assoc. Petrol. Geol.*, **45**, 1789–1799, 1961.

Joides, Ocean drilling on the continental margin, *Science*, **150**, 709–716, 1965.

Joides, The Deep Sea Drilling Project, *Trans. Am. Geophys. Union*, **48**, 817–832, 1967.

Knauss, J. A., Measurements of the Cromwell Current, *Deep-Sea Res.*, **6**, 265–286, 1960.

Knauss, J. A., Further measurements and observations on the Cromwell Current, *J. Mar. Res.*, **24**, 205–240, 1966.

LaFond, E. C., Temperature structure in the sea, in *Encyclopedia of Oceanography*, ed. by R. W. Fairbridge, 758–763, Reinhold Publ. Co., New York, 1966.

Lek, L. v., Die Ergebnisse der Strom- und Serienmessungen, *"Snellius" Expedition Reports*, Vol. II, Part 3, 169 pp., 1938.

Peterson, M. N. A., N. T. Edgar, C. von der Borch, M. B. Cita, S. Gartner, R. Goll, and C. Nigrini, *Initial Reports of the Deep Sea Drilling Project*, Vol. II, 501 pp., U.S. Government Printing Office, Washington, D.C., 1970.

Pillsbury, J. E., Report on deep-sea current work in the Gulf Stream, *Report of Supt. of U.S. Coast and Geodetic Survey*, 495–501, Washington, D.C., 1886.

Pillsbury, J. E., Report of Gulf Stream explorations, observations of currents, *Report of Supt. of U.S. Coast and Geodetic Survey*, 54–57, 281–290, Washington, D.C., 1887.

Pillsbury, J. E., Gulf Stream explorations—observations of currents 1887, *Report of Supt. of U.S. Coast and Geodetic Survey*, 54–57, 173–184, Washington, D.C., 1888.

Pillsbury, J. E., Gulf Stream explorations—observations of currents 1888, *Report of Supt. of U.S. Coast and Geodetic Survey*, 467–477, Washington, D.C., 1890.

Pillsbury, J. E., The Gulf Stream—a description of the methods employed in the investigation and the results of the research, *Report of Supt. of U.S. Coast and Geodetic Survey*, 461–620, Washington, D.C., 1891.

Richardson, W. S., P. B. Stimson, and C. H. Wilkins, Current measurements from moored buoys, *Deep-Sea Res.*, **10**, 369–388, 1963.

Sandström, J.W. and B. Helland-Hansen, Über die Berechnung von Meeresströmungen, *Report on Norwegian Fishery—and Marine Investigations, Bergen,* **2** (4), 43 pp., 1903.

Schlee, J. and R. Gerard, Cruise report and preliminary core log, *M/V Caldrill I,* 17 April to 17 May 1965, Joides Blake Panel Report, 64 pp., 1965.

Schubert, O. v., Ergebnisse der Strommessungen und der ozeanographischen Serienmessungen auf den beiden Ankerstationen der zweiten Teilfahrt, *Ann, Hydrogr., Berlin,* Jan. Beiheft, 74 pp., 1944.

Sverdrup, H. U., The influence of bottom topography on ocean currents, *Applied Mechanics,* Th. von Karman Anniv. Vol., 66–75, 1941.

Swallow, J. C., A neutral-buoyancy float for measuring deep current, *Deep-Sea Res.,* **3**, 74–81, 1955.

Thorade, H., Die Gezeitenwelle des Atlantischen Ozeans, *Ann. Hydrogr., Berlin,* **62**, Heft 1, 1–7, 1934.

Tomczak, M., Jr., Schwankungen von Schichtung und Strömung im westafrikanischen Auftriebsgebiet während der "Deutschen Nordatlantischen Expedition" 1937, *"Meteor" Forschungsergebnisse,* Reihe A, No. 7, 109 pp., 1970.

Wüst, G., Florida und Antillenstrom, eine hydrodynamische Untersuchung, *Verhandl. Inst. Meer,* Berlin, N.F.A., Vol. 12, 1924.

Wüst, G., *Stratification and Circulation in the Antillean-Caribbean Basins,* Part I, Spreading and mixing of the water types with an oceanographic atlas; 201 pp., Columbia University Press, New York, 1964.

An Inexpensive Method
for Measuring Deep Ocean Bed Currents

JAMES N. CARRUTHERS

National Institute of Oceanography Wormley, Godalming
Surrey England

Abstract An account is given of an inexpensive device with which single but easily-repeatable measurements of abyssal bed current can be made. No camera is used and risk of loss is negligible. The tests so far made are reported. There are no consumables.

PREAMBLE

Not surprisingly, this Festschrift honouring a man so full of years and so rich in achievement as Georg Wüst, contains papers devoted to many topics. The present contributor who greatly appreciates the privilege of having been invited to write an article for it, has known Wüst for so long and so closely over some years of his life, as to think that it would be regrettable not to touch upon the human side so to speak in explanation of the choice of topic.

The month of June 1928 was a very good time to have first made close acquaintance with Wüst in Berlin. It was the time of his 38th birthday and he was extremely but very happily busy in his capacity as General Secretary of the Gesellschaft für Erdkunde zu Berlin which, in the preceding month, had celebrated its hundredth anniversary. Wüst was surrounded by congratulatory addresses and elegant vellums sent in to the Gesellschaft from numerous learned societies of many nations, and, of course, the triumphs of the great Meteor Expedition of 1925–1927 were much to the fore. Wüst had been awarded the Karl Ritter Silver Medal doubtless in tribute to the scientific leadership of the expedition which he had so efficiently exercised since his illustrious mentor and brother-in-law Alfred Merz (ten years his senior) had left the ship in June 1925 owing to illness.

6 Gordon II

I remember very clearly the subject of my first conversation with Dr. Wüst at the Institut für Meereskunde. He expressed surprise at so little notice having been taken of the fact that Merz and he had (in 1921) demolished that long-lasting Lenz picture of the Atlantic circulation which showed an all-depths symmetry about the equator and the rise to surface of cold abyssal waters there. Ever an ardent admirer of the Challenger Expedition, Wüst had studied in detail the behaviour of the Casella maximum and minimum thermometers and had decided that their temperature reactions were slow enough for them to reveal inversions in the depths. Basing themselves upon a meticulous re-working of all existing oceanographical data of consequence, Merz and he had produced their well-known picture of the circulation of the waters of the Atlantic which appeared in colour as such a striking diagram in the Jubilee Volume of the Gesellschaft.

Thereafter Wüst's interest in the deep circulation never flagged, as a series of fine papers over many years amply attest. The culmination was his great report of 1933 issued in the Meteor Report series and its follow-up in 1957—of which latter he tabled a very informative summary at the Toronto meeting of the I.U.G.G. in 1957. Wüst was ever hopeful that the time would come when some of his computed velocities in the depths might be checked by actual instrumental observations, and one remembers his delight when it was made known that Swallow Float travels at 2800 metres depth in latitude 33° N and longitude 76° W, had revealed a southgoing current of as much as 17.4 cm/sec. This was in 1957 the year in which Wüst had published his computed value of current speed in the axis of the North Atlantic Deep Current at the depth of 3500 metres at 7° N latitude at the western margin of the West Atlantic Trough, as being 17.4 cm/sec towards SSE. Because Wüst had inferred the absence in the eastern trough of the Atlantic of "a continual meridional deep water circulation with measurable velocities", it has seemed fitting for the present writer to devote his contribution to this Festschrift to a description of a simple and instrumentally-cheap way of making single but easily- repeatable observations of abyssal bottom currents. Because the instrument in question has so far been tested on deep bottoms only in the eastern part of the North Atlantic, the relevance is to Wüst's inference just referred to.

THE INSTRUMENT—THE ABYSSAL PISA

The instrument to be briefly described is the latest of a family of buoyant bottom-tethered uphanging negative pendulums to which the convenient nickname "Pisa" has been given because, like the famous leaning tower, they incline from the vertical. This they do in the presence of a current, and each has within it a means for indicating the speed and direction of the current which tilts it. In preparation for the writing of a paper in which the various Pisa instruments will be described in detail and illustrated by workshop drawings, a large collection has been made of published papers by many authors who have concerned themselves with deep near-bottom currents. For the most part, what has been written about currents really close to abyssal bottoms has dealt with inferences made from photographs taken with deep-sea cameras. Sometimes the inferences have been made from the bending-over of stalked bottom growths; sometimes from the movements of ascending and descending jets of dye; sometimes from the dispersal of disturbed clouds of sediment, and often from presumed erosion tails behind boulders and the like. Other information has come from obser-vations of various kinds made visually from submersibles. We may mention also the ovservations made by Thorndike and others using nephelometers to study the nepheloid layer above the ocean bottom, and the interesting findings of Lafond who viewed by underwater television the movements of white nylon yarn streamers against a framework of reference. The deep instrumental observations reported by J. A. Knauss also call for mention here, but they related to a level some ten metres above bottom. Much in-formation of first-class value on abyssal bed currents is expected from the remarkable capsule developed by F. E. Snodgrass but, if thought be given to existing instrumental data on current speeds and directions really close to very deep bottoms, the amount of information is sparse indeed!

The present writer has nothing to write about which can vie even remotely with the sophisticated off-bottom current measuring instruments existing today, but he has long held the opinion that an useful purpose could be served by simple instruments which involve only negligible cost, involve no cameras, and can be operated with trifling risk of loss. Though those which he has made can measure current speed and direction very close indeed to abyssal bottoms, they have the serious shortcoming that they do not record but serve only to give single observations. Their justification to him has lain in what may be called their survey function—their ability to reveal what the abyssal bed current *can* be at many places along the course of an exploring vessel which either has no deep-sea camera or no time or dis-position to use one, but which could spare time at interesting spots to make single on-bottom observations. An an example of a series of single Pisa observations of deep bed current, those made by H. M. S. 'Vidal' at 13 loca-tions in the Gulf of Cadiz in June 1964 and published by Heezen and John-son in Lamont Geological Observatory Contribution Number 1119 can be referred to.

The simple instrument under present description (Abyssal Pisa—"A. P."

from now on) is entirely mechanical, whereas its forerunners made use of
the earlier jelly bottle principle still much in use for observing at lesser
depths. Years ago the pisas were used from drifting ships and were therefore equipped with a pay-out line which became operative at bottom when
they had been placed on the seabed. This was of course to compensate for
ship drift, but the method was troublesome and has now been abandoned
in favour of use from a ship which can keep her position over the ground
by use of main engines and a bowthruster.

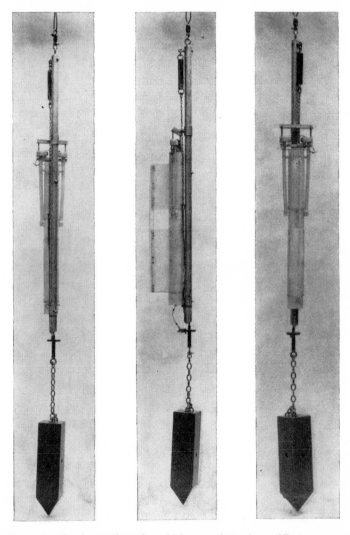

Figure 1   The descending Abyssal Pisa seen from three different aspects

It hardly needs remarking that an excellent observing platform for the use of the instrument would be a dynamically-positioned boring ship such as Glomar Challenger. To have an observer aboard such a ship who could make many repeated A.P. observations during her stay of a few days at each of her many drilling locations, would seem ideal, and it would certainly lead to useful knowledge of abyssal bed currents being obtained at least imaginable cost.

DESCRIPTION

We necessarily commence with the remark that some of our dimensions may seem awkward because of conversion into metric values.

A main part of the instrument is a tube of low density polythene one metre long and of wall thickness 6.4 mm. This of course would float in water. It is closed at what is to be the bottom end by a welded-in polythene disc having at its external centre an eye for the attachment of a tether. This latter is about 25 cm long and is made of stranded steel wire encased in a stocking of vinyl tubing. It has a small brass shackle at each end—one end being shackled into the basal eye of the tube. Around the tube is fixed a streamlined casing of thin sheet polythene. The line from the centre of the tube to the closed edge of the shaped casing (the fin) will, of course, be the fore-and-aft direction of the instrument when in use. About half-way down from the open top of the tube an arm which is virtually a springy lever of stout polythene rod, is hinged to each of the thwartships sides. These reach up slightly higher than the tube top, and, at their upper extremities, they are pulled together by a constant tension spring of non-rust steel which, in the absence of anything else, would constrain the arms to lie snug to the tube. Inside the tube just below its open top, two discs of hard plastic disposed in the fore-and-aft line face each other. These discs which are 8 cm in diameter and 1 cm thick, are borne upon circular rods which come into the tube through easy-fit round holes cut into it. Externally to the tube, these rods are welded into the lever arms at right angles. It results that movements of the arms will push the rods inwards or pull them outwards. The discs are fastened at their centres to the inwards ends of the rods and at right angles to them so that their faces are well apart or close according as the arms are pulled apart or allowed to snug the tube under the tension of the spring. One of the discs which is fixed non-rotational on its rod end, is covered with a circular pad of thick rubber. The other one which can rotate freely on its rod end is graduated into degrees. This graduated disc has a cavity cut into it well off centre, and in this cavity there is a pivotted disc compass strongly magnetised across its diameter. The compass is marked by

directional paint stripes—red for north and blue for south, and the base of
its pivot is set into a small brass slug at the bottom of the cavity to provide
weight. The edge of the compass projects well out from its disc so that,
when the discs are close together, the compass is jammed tight. It will be

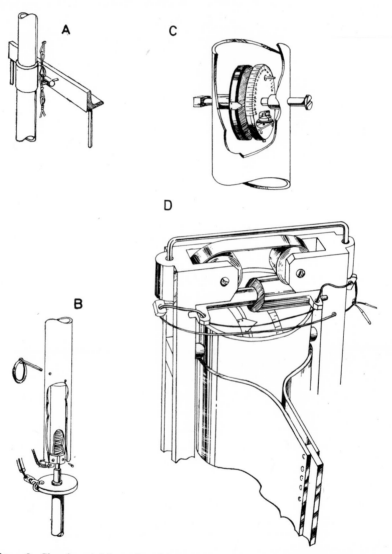

Figure 2 Showing at A how the release horns are attached to the chain; at B the
attachments at the bottom of the lowering gear with the safety pin on the left; at C
the compass disc and the locking disc—and at D the top of the instrument to
illustrate the streamlined body, the piston insert, the constant tension closure
spring, and the two discs

obvious that the disc containing the compass is a sort of circular pendulum in that the weight of the compass and its pivot baseplate act to ensure that the compass is always downwards in the tube when the discs are apart no matter how the tube be sloped—open end upwards of course. To minimise the side pressure on the compass pivot when the discs are closed together, a pointed pin projects out from the compass disc. This, which stabs into the rubber pad of the fixed disc, not only controls the pressure but locks the compass disc firmly to preserve its inclination.

To prepare for use, the arm tops are pulled apart against the tension of the spring and a water-filled plastic piston provided with a time-adjustable bleed leak is inserted horizontally between the arm tops in a thwartships direction. The two ends of this piston are held on thumbscrew bolts set through the arms, but a loss-preventer string is fitted to prevent loss of the piston which becomes loose after closure. When the filled piston is inserted, it is kept free of closure pressure initially by a metal restrainer bar. This is a metal rod bent downwards at its two ends into "horns" which are pushed into holes in the arm tops. If this bar were snatched upwards, the spring would squeeze the piston to the end that, after the set timing had elapsed (usually about $3\frac{1}{2}$ minutes) the discs would come together to lock the compass disc in respect both of inclination and direction. Of course, the inclination of the compass disc measures the tilt of the tube, and the amount of tilt is convertible into current speed from a graph established by appropriate calibrations carried out in a circulating channel or in a towing tank. The inclination is read from a pointer leading out from the edge of the fixed disc towards the graduated rim of the rotating compass disc.

As so far described, if the streamlined instrument were tethered to a weight on the bed of a stream deep enough to cover it, and the horned restrainer bar were snatched out, the whole thing would tilt in accordance with current speed. When the piston had been squeezed in by spring pressure, the discs would come together and, on retrieval, one would be able to read the inclination (*i.e.* speed) and the direction of flow. To use the instrument on deep bottoms it is necessary to arrange that the A.P. becomes freely tethered when it reaches the ocean bed, and that the snatch-out of the restrainer bar be accomplished when bottom is reached. It must not matter how long the operation of lowering lasts; nothing must happen during descent.

The method of working adopted is as follows:

Firstly, a length of 1.8 m is cut from an aluminium alloy tube 5 cm in diameter. This is provided (at what is to be its top) with a large eye for the attachment of the lowering line. This latter will usually be the standard hydrowire. The eye will of course be connected to the lowering line via a swivel. Transversely through the top of the tube an eyebolt is fixed to provide an attachment for a strong coil spring whose function is to go down stretched and to effect the snatch-out of the restrainer bar on bottom touch. The spring is attached to the eyebolt at the top and then, to the lower end of it a length of jumbo chain (55 cm) is shackled on. To the bottom end of this chain is attached a length of strong cord having at its end a domino of teflon with a smooth vertical hole through it. Up into the alloy tube at its bottom end runs a round rod furnished with a bottom eye into which will be shackled the eventual sinker weight. The projection of this rod below the tube varies a little dependent upon whether it is or is not pulled strongly downwards to compress a spring inside the tube. When it is pulled down strongly a short smooth finger emerges alongside it at the bottom of the alloy tube; otherwise the finger withdraws completely into the tube bottom. When the rod is pulled down against the resistance of the internal spring, a pin (hereafter referred to as the safety pin) can be passed through the tube to hold the rod at its full projection with the finger poking out at base. Having pulled down on the side chain above referred to, the teflon domino at the end of the cord is placed over the protruding finger so that the latter is seated in the smooth hole of the domino.

As so far described, if we hung the alloy tube from a hook in the ceiling and then fastened a very heavy weight to the bottom eye, we could withdraw the safety pin to leave the side chain and cord held down against the tension of the coil spring at the top of the tube. Actually we should have a short length of chain between the sinker weight and the bottom eye. If now, with the safety pin out, we could lower the assembly to let the weight rest on the floor, the finger would draw in, the domino would slip off, and the chain and cord would jump upwards. That is the action which takes place in actual use of the instrument.

Sliding on the alloy tube is a free collar which carries a bar with two downpointing horns like the restrainer bar already referred to. We set the instrument as described already i.e. with the filled piston in position and the restrainer bar holding the arms apart. Then we insert wooden wedges horizontally between the arms and the polythene tube to keep the piston at full extent when we necessarily remove the restrainer bar. We next put the pisa alongside the alloy tube whilst this latter is in the set situation i.e. with its spring extended, the teflon domino over the finger, and the safety pin in place. We then shackle the free end of the tether to a little shackle carried at the periphery of a spinning wheel of tufnol (9 cm diameter and 3.2 mm thick) which can turn freely on the projecting part of the rod from whose end the sinker will be hung. This wheel is held in position by means of a short sleeve of tufnol tubing both above and below it.

At the base of the pisa is a little loop of red plastic twine down through which can be passed a stout smooth skewer to seat into an eye on the side of the long alloy tube. This skewer is at the end of a string whose top is fastened above to the chain at such length that it is just taut when the gear is set. This device holds the bottom of the pisa in position but does so in a way which allows easy release. Then the top of the pisa is pressed against the alloy tube whilst its arms are held apart by the side wedges. The horns of the sliding collar are then placed in position—sinking well down into holes provided in the arm tops. Then the collar is fixed to the chain by dint of passing through a link of the latter a thumbscrew-headed bolt which is screwed into the loose collar. Then the wooden wedges are withdrawn to leave the arm aperture maintained by the horned bar.

The cocked assembly is hung from the end of the hydrowire via the swivel and swung outboard with the heavy weight (about 90 kg) shackled to the sinker eye via a length of about 60 cm of chain. Once the weight is hanging free from the bottom of the assembly the safety pin is withdrawn.

The gear is now ready for lowering. When the heavy sinker touches bottom the finger pulls in, the chain jumps upwards, the restrainer bar withdraws and the bottom restraint of the pisa against the alloy tube is removed by the withdrawal of the skewer. Thereafter, with a little more veering of the wire, the alloy tube falls over to lie on the seabed. Because of its buoyancy the pisa "floats" upwards without risk of being trapped because its tether can spin round on the rotating tufnol wheel. It results that the pisa stands erect without restraint in still water or leans over under the influence of any existing bottom current. After being left on the seabed for the few minutes ordained by the piston timing, the whole assembly is winched back to surface. Used on bottoms sounding no more than a thousand metres or so, bottom touch can be detected by watching the dynamometer springs. In such case the extra wire necessary can be paid out whilst the ship is held in position by main engines and bowthruster. Where the depth is very great, the practice has been to secure a pinger on to the hydrowire about 100 m above the A.P. Then, when bottom touch has been revealed by the pinger, the necessary wire to bottom the A.P. and obviate dragging is paid out, and the ship kept in position for the requisite few minutes before hauling in.

Because the A.P. has not been in its really-finished state for very long, it has not yet been possible to make many observations with it on really deep bottoms.

Working from H.M.S. Hydra on 20th November 1967, I used the A.P. at 5000 m depth at the position 44° 11′ N–10° 50′ W and there measured a bottom current of about 4 cm/sec towards S.W. The whole operation took about 3 hours and a check was provided by another kind of pisa used simultaneously. Since that time the piston has been much improved and simplified, and recently my colleague Mr. Crease put the A.P. down to 4700 m depth from the R.R.S. Discovery at the position 51° N–17° W which location is some 480 miles west of the Bristol Channel. His result was no detectable current.

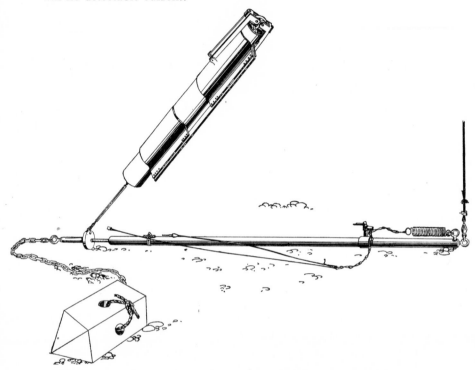

Figure 3   Showing the Abyssal Pisa in the operating situation on bottom. The vertical line on the right is of course the end of the hydrowire

It is naturally hoped that opportunities for many more measurements will soon present themselves.

THE CALIBRATION

When the calibrations were made they were of course carried out in fresh water. Computations were made to extend the results to seawater, and curves were made for use in ambiences of the following densities:

1028, 1038, 1048

Of course, if desired, one could follow a practice made on an earlier occasion in order to obviate the need for the transference computations. The buoyance of the A.P. could be measured when submerged in tanks of salt waters of various densities made up in the laboratory. That done, the instrument could have its buoyancy "doctored" to suit the buoyancies it would possess in those waters—and then re-calibrated in fresh water with the increased buoyancies.

It was desirable to allow for any change in buoyancy of the A.P. on account of compression when used at really great depths. I had no means of making this determination and so had recourse to the good offices of Mr. Gilbert Jaffe who heads the National Instrumentation Center at the U.S. Naval Oceanographic Office. He very kindly made the necessary determinations with the result that the calibration curves issued with the A.P. carry an indication of the allowances which have to be made to cater for the changes in buoyancy occasioned by deep submergence. We do not reproduce the calibration curves here, but that for an ambience of 1048 water can easily be plotted from the following points where T is the tilt angle in degrees and S the speed in knots:

$$T — 15 \quad 35 \quad 48 \quad 57 \quad 64 \quad 68 \quad 71 \quad 74 \quad 76 \quad 77$$

$$S — 0.1 \quad 0.2 \quad 0.3 \quad 0.4 \quad 0.5 \quad 0.6 \quad 0.7 \quad 0.8 \quad 0.9 \quad 1.0$$

The allowances to be made for compressibility of the polythene are as follows where D is the submergence depth in metres and R the percentage by which the graphed speed is to be reduced:

D	1000	2000	3000	4000	5000	6000	7000
R	1.5	3.1	4.3	5.6	6.7	7.7	8.5

Acknowledgements

For the constructional work I am indebted to my colleagues Messrs. R.E. Potter and D. Bookham; for the diagrams to Mr. N.R. Satchel and Miss P.E. Williamson of N.I.O.; for calibration facilities to Professor J.H. Preston whose circulating channel at Liverpool University I used, and for the compressibility tests to Mr. G. Jaffe as already mentioned.

The Effect of the Reykjanes Ridge on the Flow of Water Above 2000 Meters*

MARCUS G. LANGSETH, JR.

Lamont-Doherty Geological Observatory of Columbia University
Palisades, New York

DON BOYER

University of Delaware, Newark, Delaware

Abstract   Drift and set of the research ship *Vema* during a detailed survey of the Reykjanes Ridge revealed a surface current pattern with flow to the southwest over the eastern flank of the ridge and northeast over the western flank. The flow is essentially parallel with the isobaths of the plunging ridge. Dynamic calculations using hydrographic data indicates that geostrophic flow above 1500 m sometimes follows the same pattern but the pattern is not a constant feature of the area.

Experiments carried out in the laboratory modeling flow over a plunging ridge on a rotating frame demonstrate that such a pattern can occur when a deep westward flowing current encounters a southward dipping ridge crest. The model also clearly shows the boundary flow at the sea floor over the ridge. This flow is away from the axis and may in part explain the thick deposits of sediments found at the base of the ridge.

I   INTRODUCTION

The Reykjanes Ridge is a section of the Mid-Atlantic Ridge system that extends southwestward from the southern continental slope of Iceland. The elevations near the axis of the ridge decrease in a regular way away from the continental shelf so that on a gross scale the Reykjanes Ridge forms a "snout" extending southwestward from the tip of the Reykjanes Peninsula. The descriptive term "snout" was used by Cooper (1955) in discussing the influence of this ridge on the flow of deep water in the Northeastern Atlantic

---

* Lamont-Doherty Geological Observatory Contribution Number 1606.

Basin. A detailed study of the flow of deep water in the vicinity of the Reykjanes Ridge by Worthington and Volkmann (1965) showed that the Norwegian Sea overflow water moves as a contour current along the flanks of the ridge below 1500 m. The general pattern of flow of this deep water is shown in Figure 1, which is reproduced from Worthington and Volkmann's paper.

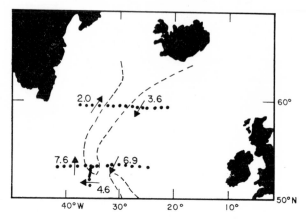

Figure 1 Direction of flow and volume transport (millions of m³/sec) of the Norwegian Sea overflow water in the vicinity of the Reykjanes Ridge. From Worthington and Volkmann (1965)

In this paper we will show evidence that the ridge also has an influence on the current pattern above 1500 m including the surface currents. Drifts of the research ship *Vema* during a geophysical survey of the Reykjanes Ridge between 60° and 63° N (Talwani *et al.*, 1970) indicated surface currents that flow predominantly southwestward along the eastern flank of the ridge and northeastward along the western flank; essentially the same direction of flow deduced for the deeper water. Hydrographic data in this area indicate that at times the flow pattern above 1500 m has the same general characteristics as that obtained from the ship's drift. Data from several years indicate the pattern of flow in this region is variable.

Model studies, that are described in this paper, demonstrate that the observed current pattern may be a natural consequence of flow over the plunging ridge on the rotating earth. In the studies, the flow directions in the Ekman boundary layer on the sloping sides of the model ridge are depicted using dye tracer techniques. The observed Ekman transports are away from the ridge axis on both sides of the ridge. These boundary flows are one mechanism whereby sediments are transported toward the base of the ridge.

II THE SURFACE CURRENT PATTERN OVER THE REYKJANES RIDGE

Surface Currents from Ship's Drift

Navigation by satellite at sea permits more accurate and more frequent determination of a ship's drift and set. If two tracks with several fixes are run in opposing directions in relatively light winds, these drifts and sets can give a good measure of the magnitude and direction of surface currents.

In August 1966 a geophysical survey was made of the crestal region of the Reykjanes Ridge. The Lamont-Doherty Geological Observatory research ship *Vema* made eleven parallel tracks across and normal to the axis of the ridge; see Figure 2a. All fixes during this survey were made with a satellite navigation system. Talwani *et al.* (1970) have demonstrated that the errors associated with these fixes are usually less than 0.1 nautical mile. While making these tracks, the *Vema* was steered on fixed headings of either 127° or 307°. Steering was by autopilot and the engine revolutions per minute were held constant so that dead reckoning positions could be determined with the greatest possible accuracy.

The apparent drift vectors between satellite fixes are shown in Figure 2. The magnitude of each vector, or drift, is calculated by dividing the distance

Table 1 Drift vectors for various track segments on the Reykjanes Ridge Survey

Track Number	Course	Column 1 Net drift vector of each track		Column 2 East segments only		Column 3 West segments only		Column 4 Net drift vector of two opposing tracks	
		drift	set	drift	set	drift	set	drift	set
1	307	0.4	153	0.15	238	0.20	003	0.07	302
2	127	0.5	330						
3	307	0.4	179	0.34	258	0.23	346	0.18	291
4	127	0.6	327						
5	307	0.5	163	0.30	215	—	—		
6	127	0.3	296						
7	307	0.3	202	0.33	236	0.13	337	0.16	288
8	307	0.3	198						
9	127	0.4	333						
10	307	0.1	293						
11	127	0.8	146						
12	037	0.2	315						

[1] The drift is in knots and the set is in degrees by the compass.

between the dead reckoning position (DR) and the satellite navigation
position (SATNAV) by the time interval between adjacent fixes. The direction of the vector, or set, is from the DR position toward the SATNAV
position. These apparent drift vectors are made up of four major components:
1) current induced drift and set, 2) wind induced drift and set, 3) dead
reckoning errors, and 4) satellite navigation errors.

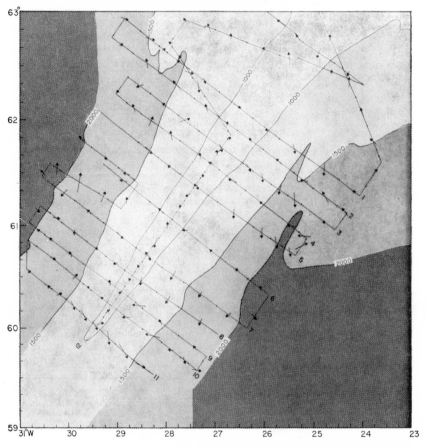

Figure 2 Tracks of the R.V. *Vema* over the Reykjanes Ridge. Satellite fixes are
shown by dots and the apparent drift vectors between fixes are indicated by arrows

In this survey most of the SATNAV fixes are separated by time intervals
greater than two hours. For intervals this large the contribution of satellite
navigation errors to the apparent drift vector is small. Dead reckoning errors,
as we shall demonstrate later, can be relatively large and some way of
eliminating these errors from the apparent drift must be found.

In their analysis of the navigation data Talwani *et al.* (1970) determined the residual drift vectors between each pair of adjacent SATNAV fixes along the tracks. The residual drift vector, $\vec{r}$, is defined as: $\vec{r} = \vec{a} - \vec{n}$, where $\vec{a}$ is the apparent drift vector and $\vec{n}$ is the net drift vector. ($\vec{n}$ is calculated using the fixes at the ends of each transverse track). The residual drift vectors are shown in Figure 3 and the net drift vectors are listed in column 1 of Table 1. This definition of residual drift eliminates dead reckoning errors that are uniform over each track, however, it also eliminates any net components of drift and set due to currents and wind. Therefore, these vectors yield information on the relative magnitude and direction of the resultant of surface currents and wind induced offsets along each track. The vectors depicted in Figure 3 reveal that the relative set caused by currents is predominantly

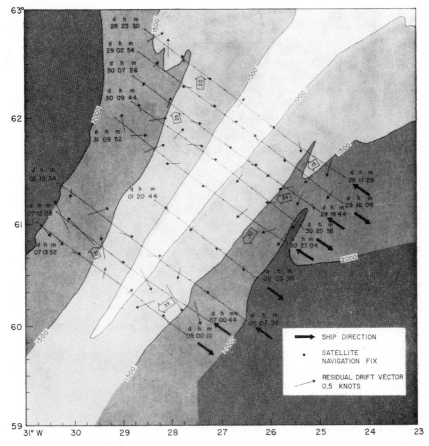

Figure 3   Residual drift vectors between satellite fixes are shown on selected tracks. The large open arrows show net drift between opposing sections of track. The speeds are given in each arrow in knots

southwest over the eastern flank of the ridge and northeast over the western flank.

The net drift vectors along the single tracks, listed in Table 1, show that the drift calculated on tracks with a 307° heading are nearly equal and of opposite set to those with a 127° heading. This is evidence that the dead reckoning errors are relatively large. The dead reckoning errors are composed of pit log and steering errors as shown in the vector diagram of Figure 4. The pit log errors are directed along the track and, in calm sea conditions, they are nearly independent of headings. The steering errors are normal to the track and generally are in the same direction relative to the heading. If we average the net drifts of two opposing tracks taking into account the time spent on each track, the dead reckoning errors tend to cancel. On the other hand,

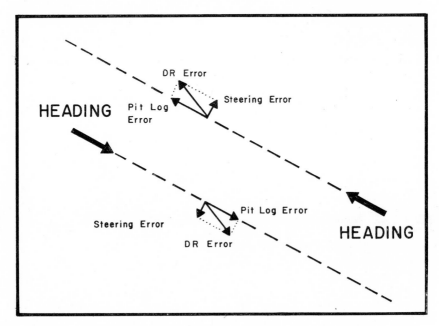

Figure 4   Schematic diagram showing dead reckoning errors on opposing tracks

wind and current induced drift and set that persist along both tracks will not cancel.

In column 4 of Table 1 we give the average drifts and sets for pairs of opposing tracks. We did not use opposing tracks 10 and 11 since during the running of these tracks, the ship encountered gale force winds. The ship's deck log contained entries of winds and seas of force 9. We also did not include tracks 5 and 6 in this calculation since most of track 6 is confined to the eastern side of the ridge. The remaining pairs give average drifts between

0.07 and 0.18 kts with a west northwesterly set. These vectors are the best indication of the net drift vector induced by wind and currents over the width of the survey area.

Gentle easterly winds affected the progress of the ship along the line of travel. These changes in speed would not contribute to the net drift vector since they are measured by the pit log. However some drift normal to the track can result from the action of wind and sea that would not be detected by the pit log and hence would be a component of the calculated net drift vectors. There is no way of determining from the existing data the drift resulting from wind and sea. We think they may account for a substantial part of the vectors listed in column 4.

Segments of opposing tracks over the same side of the ridge can be used to examine the change of direction of surface currents over the crest of the ridge indicated by the residual drift vectors. The net drift vectors for opposing track segments over the eastern flank and the western flank are given in Table 1, columns 2 and 3. The vectors for all track pairs on a given side are very consistent, and indicate a flow on the eastern side that is west southwesterly at about 0.3 knots and north northwesterly over the western flank at 0.2 knots. These results are also depicted as broad arrows in Figure 2b. Because of the consistency of the results, we feel that these vectors closely represent the average surface current velocities over the eastern and western flanks of the ridge.

The alignment of the flow pattern with the axis of the Reykjanes Ridge suggests that there is a hydrodynamic interaction between the surface currents and the upper portions of the ridge. To further illustrate the morphology of the ridge we show four transverse profiles made by the seismic reflection technique in Figure 5. These profiles show not only the topography of the sea floor but the basement surface beneath and some layering within the sediment (Ewing and Zaunere, 1964). The average depth near the axis is about 700 m at the northern end of the survey area, profile 1. In cross section the central crestal regions in profiles 1 and 3 appear as elevated plateaus relative to the flanks of the ridge. In the southern profiles the ridge has a more triangular shape in profile and the average depth at the axis is about 1100 m. Also see the generalized contours in Figure 2, which are drawn from survey data.

To encounter the ridge the currents observed at the surface must extend to depths of about 1500 m. We will next examine hydrographic data in this area to determine the pattern of flow with depth. First we shall define the water masses found over the Reykjanes Ridge based on temperature and salinity characteristics, and then describe the flow of water above 1500 m based on geostrophic calculations.

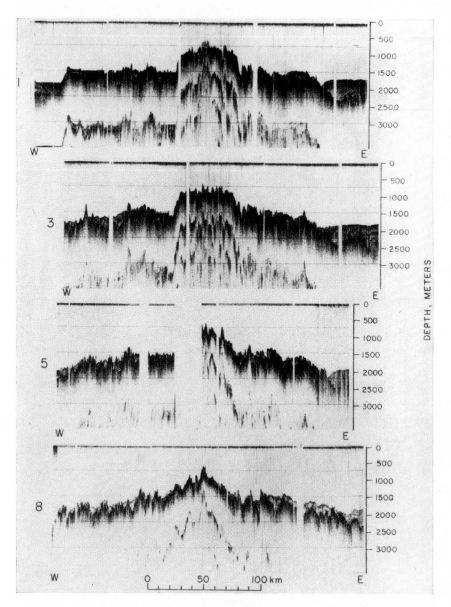

Figure 5 Four transverse profiles of the Reykjanes Ridge made by the seismic reflection technique (Ewing and Zaunere, 1964). Numbers on the left hand side of each profile corresponds to numbers in Figure 2

Water Masses Over the Ridge

Nearly one hundred hydrographic stations have been taken in the vicinity of the survey area over the past 40 years. Plots of temperature versus salinity for three typical stations taken in April 1962 (Worthington and Volkmann, 1965) are shown in Figure 6. Note we used stations from the east and west flanks of the ridge as well as over the crest. The characteristics of the water over the ridge are formed by the mixture of three source waters: 1) North Atlantic water (NA), 2) Irminger Sea water (IR), and 3) Norwegian Sea overflow water (NS). The source characteristics of NA and IR water defined by Dietrich (1957) and Lee (1967) are both plotted on the diagram. The Norwegian Sea water characteristics are given as $-0.4°C$ and $34.92°/_{oo}$ (Lee, 1967).

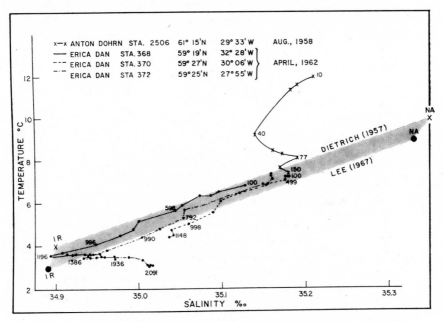

Figure 6 Water characteristics over the Reykjanes Ridge plotted on a temperature versus salinity diagram. Characteristics of the (NA and IR) source waters given by Dietrich are given as X's and those by Lee as dots

Points between 100 m and about 1000 m lie along a mixing line between NA water and IR water. Water with characteristics near those of the NA water are found near the core of the Northeast Atlantic surface current that flows generally northward in the Northeastern Basin of the Atlantic. Near the ridge this water mixes with the less saline and colder IR water with depth. The IR water is found in a wide range of depths, 50 to 1500 m, in the central parts of the Irminger Sea. Depths given along the curves in Figure 6 indicate that this mixed water mass extends to depths of 1200 to 1400 m.

During the summer the upper 100 m of the water is warmed reaching about 12 °C at the surface. A typical summer T-S profile above 150 m taken by the Anton Dohrn in August 1958 is shown in Figure 6. Below 150 m the summer water curves merge with the winter curves from the *Erika Dan*. Observed lateral variations of properties in the warmed summer layer do not have a significant effect on the dynamic topography and consequently have a relatively small effect on the general surface currents. The geostrophic current pattern in the NA-IR mixed water is in the main determined by the field of mass between 200 and 1400 m.

At depths of 1200 to 1400 m the T-S curves turn away from the NA-IR mixing line toward higher salinities and lower temperatures. The resulting salinity minimum marks the upper boundary of Norwegian Sea overflow water. The salinity of the overflow water is often found to be greater than $35^o/_{oo}$ presumably as a result of entrainment of NA water as it flows down and along the southern slopes of the Faeroe–Iceland Ridge.

In terms of *sigma t* the density of the water near the boundary between the NA-IR water and the deeper water is very near 27.80. The salinity minimum or the 27.8 isopycnal have been used as the level of no motion for dynamic calculation in this area (Worthington and Volkman, 1965, and Dietrich, 1957). In the calculations presented in the next section we will use the intermediate salinity minimum as the reference level.

The Surface Circulation in 1955

Dietrich (1957) deduced the pattern of surface flow based on dynamic topography of the sea surface during the summer of 1955. His results show that a northern branch of the Gulf Stream system, the Northeast Atlantic Current, is involved in the flow over the ridge. The general directions are illustrated in Figure 7 (upper left). The current flows northeastward parallel to the axis of the ridge, but north of 60° N the current breaks into two branches. The eastern branch swings abruptly toward the east and passes along the Faeroe–Iceland Ridge. The western branch or Irminger Current turns northward toward the Denmark Straits. Over the shallower parts of the ridge, between 62° and 64° N the dynamic topography suggests two weakly developed gyrals. Over the ridge axis between 60° and 62° N the flow is relatively sluggish and toward the northeast. The branching of the current indicates that the plunging crest of the ridge does influence the current flow north of 60° N. However, the flow pattern based on dynamic topography in June 1955 is quite different than that deduced from ship's drift in August 1966. In particular, the southward flow over the eastern flank is missing.

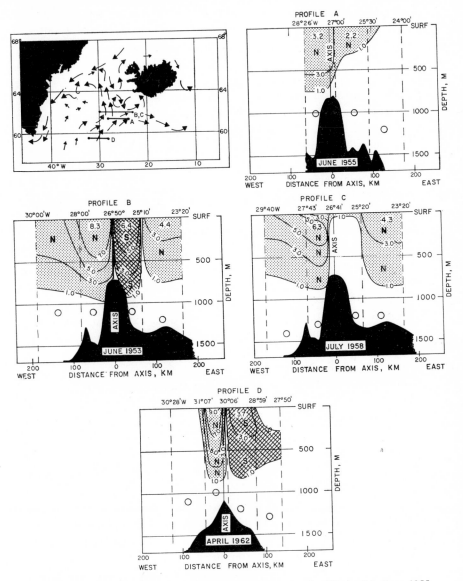

Figure 7 (Upper Left) The surface current pattern south of Iceland in June 1955 modified from Dietrich (1957). (Upper right) North and south components of currents based on dynamic calculations using Dietrich's data along profile A. The contours are labeled in cm/sec and the reference level is shown as open circles. Vertical dashed lines are drawn below station locations. Lower frames show the currents normal to selected profiles at three different times

The hydrographic data from other years show that the pattern of geostrophic flow is highly variable. At times it closely resembles the pattern observed by Dietrich and at other times that observed during the 1966 survey.

Geostrophic Flows Over the Ridge in 1953, 1958, and 1962

In Figure 7 we show the results of geostrophic calculation for four transverse sections over the ridge. We have used data from the Danish ship *Dana*, profiles B and C (see for example Hermann, 1954), and from the *Erika Dan* profile D (Worthington and Volkmann, 1965). Profile A is one of the transverse profiles made during the June 1955 survey reported by Dietrich. Three other profiles were computed that showed a northward flow everywhere over the ridge, but are not shown in the figure. Profiles B, C, and D show north-south components of geostrophic flow that are similar to those determined from ship's drift. In profile C no southward component was observed but the relative pattern of flow is the same. The southward flowing water is confined to a zone about 100 km wide on the eastern side of the ridge and is bounded on both sides by currents with northward components.

One artifice that was used in these calculations should be mentioned. The hydrographic stations at the crest of the ridge often extend to only 600 m making dynamic calculations below this depth impossible without some reasonable way to extrapolate characteristics to greater depth. Since we were interested in the flow of currents adjacent to the ridge crest and the level of no motion is deeper than the ridge crest, we resorted to interpolating the lines of equal *sigma t* between stations either side of the crest through the ridge. We used the trend of the isopycnals in the water above the crest to make these interpolations.

We do not know of any hydrographic stations that were taken during the summer of 1966 in the area of our survey; however, continuous vertical profiles of temperature were measured in conjunction with geothermal stations (Talwani *et al.*, 1970). These stations were over the western flank. Isotherms drawn to the temperature points dip down steeply toward the crest of the ridge. Using the T-S diagram it is possible to crudely estimate the *sigma t*'s corresponding to these temperatures. Such estimates show that the isopycnals also dip toward the ridge axis indicating a relatively strong northward geostrophic flow above the level of no motion over the western flank at the time the survey was made.

The surface current pattern that involves southerly flow over the eastern flank of the ridge and northerly flow over the western flank occurs only at times. Nonetheless its existence must be fairly frequent since it was indicated in about a third of the hydrographic sections made across the ridge and also in the summer of 1966. Such a pattern could result from a greater development of the small gyrals on the shelf south of Iceland. For example, if the counterflow of the Irminger Current southward along the Greenland–Iceland Ridge suggested by Dietrich (1957) were stronger, it might result in a southward flowing current extending as far north as 60° N.

We think it more likely; however, that the diversion of a westward flowing current along the contours results when it encounters a plunging ridge. We have been able to demonstrate by means of model experiments that it is possible to produce just such a pattern by a current flowing over a plunging ridge on a rotating platform. A description of these model experiments is given in the next section.

III LABORATORY EXPERIMENTS OF FLOW OVER A PLUNGING RIDGE

Rotating Water Tunnel

One of us, Boyer (1971), has recently developed a rotating water tunnel which can be used to investigate the effects of bottom topography on baratropic flows in rotating systems. The basic capability of the tunnel is that it can provide a uniform rectilinear flow relative to an observer on rotating frame. Since the tunnel is discussed in some detail in Boyer (1971), only a brief description of the apparatus will be given here. The tunnel is a channel of rectangular cross-section which rotates at a constant angular velocity, ω,

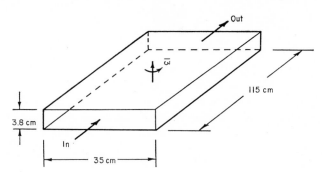

Figure 8   Dimensions and orientation of the rotating water tunnel

about an axis perpendicular to its own axis (Figure 8). Water is recirculated through the tunnel and, by adjusting the entrance and exit conditions, it is possible to obtain a uniform flow in the central portion of the channel; i.e., outside of boundary layers on the channel walls. The boundary layers on the horizontal surfaces are Ekman layers whereas those on the vertical walls have a complex structure which is not well understood, at least for the range of parameters considered here. However, flow in the area of interest, near the central portion of the tunnel, is far enough from the side walls so that these vertical layers need not be considered.

The topographies to be investigated are placed in the channel in symmetric pairs as indicated in Figure 9. For such a symmetric system, the midplane, $z = 0$, is a surface on which the vertical velocity component and shear stress are identically zero. These are the approximate free surface conditions for a large scale ocean current system with negligible wind shear stresses. We restrict our attention to $z \leq 0$ in the laboratory flow. Taking the horizontal length scale, D, and the vertical scale, H, to be independent, the flow characteristics of the system depend on three dimensionless numbers; the Rossby number, $Ro = U/2\omega D$, the Ekman number $E = \nu/2\omega D^2$ and the aspect ratio H/D where H is the water depth, D is the topography width, U is the free stream speed, and ν, the kinematic viscosity (Figure 9).

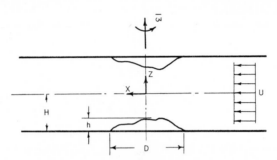

Figure 9 Vertical section of the water tunnel showing the symmetrical placement of topographies

In Boyer (1971) the flow over a long ridge of constant cross-section was investigated both analytically and experimentally. The photograph in Figure 10 is taken from this paper. It depicts the midplane streamlines of the flow over a triangular ridge for an experiment in which the free stream flow is normal to the ridge axis. We first note that there is no upstream influence, i.e., the fluid particles are not deflected until they are directly above the ridge. In crossing the ridge the streamlines are deflected to the right and far downstream they again become parallel to their upstream directions. The net effect is a finite shift to the right for each streamline. Although this experiment and others, reported in Boyer (1971), were for ridges of very small height to width ratios, other recent experiments with order unity aspect ratios indicate a similar flow behavior. The basic difference is that the net displacements are much larger for the taller ridges. It should also be noted that experiments were conducted for which the free stream flow was oblique to the ridge axis.

The resulting flow patterns again were qualitatively similar, in that the streamlines were deflected toward the right and experienced a net rightward displacement far downstream of the ridge. These flow characteristics should be contrasted to the flow over a plunging ridge discussed below.

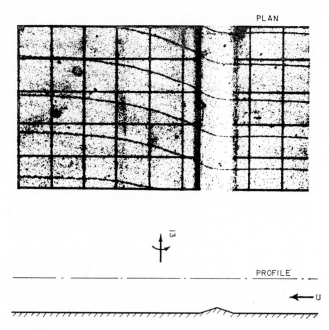

Figure 10   Mid-plane streamlines for the flow over a triangular ridge of constant cross-section ($H/D = 0.75$, $h/D = 0.063$, $E = 8.8 \times 10^{-4}$ and $Ro = 7.9 \times 10^{-2}$)

There are many difficulties involved in the attempt to relate laboratory flows to the real ocean. To date it has not been possible to resolve these problems in a satisfactory way. In fact, it is unlikely that laboratory modeling of such flows can ever approach the preciseness of, say, wind tunnel testing in aerodynamics. As such, any comparison of laboratory flows to currents in the real ocean must be made with caution.

The laboratory flows considered here are restricted to free stream speeds in the range $0.1 < U < 1.0$ cm/sec and rotation rates in the range $0.5 < \omega < 2.0$ rad/sec. For the channel dimensions indicated in Figure 8, the free stream flows are thus laminar. Ocean currents are, of course, turbulent and one is thus faced with simulating a turbulent flow with a laminar one. By introducing constant eddy viscosity coefficients into the equations of motion for the ocean and by making some rather restrictive assumptions on the magnitudes of these coefficients, it is possible to relate the ocean and laboratory flows; see Boyer and Guala (1972). Since the comparisons between the laboratory model and ocean prototype made below are purely qualitative, however, this approach is not used here.

In the laboratory flow the rotation rate, ω, of course does not vary in the horizontal plane. For the real ocean, on the other hand, the vertical component of the earth's rotation varies with latitude. These so-called β-effects are thus not modeled in the present laboratory system.

It is not possible to imcorporate geometric scaling of ocean floor topographic features into the laboratory model. For features such as mid-ocean ridges, for example, the aspect ratio is about 5×10^{-3} whereas in the laboratory the aspect ratios are of order unity. Despite this vertical exaggeration in the laboratory model we think the flow characteristics obtained will qualitatively represent those in the real ocean. Perhaps introducing some vertical distortion is justified since in both the real ocean and the model of the Reykjanes Ridge, the ratio of the Ekman layer thickness to the water depth is approximately the same.

It should also be noted that stratification in the ocean is most certainly important in determining any particular current system. Since stratification is completely neglected in our modeling, it is expected that the topographic effects observed in the laboratory flow will be exaggerated.

Flow Over a Plunging Ridge

We return to our consideration of the Reykjanes Ridge. It is desired to construct a laboratory model which, will demonstrate some of the effects deduced from the ship drift and hydrographic data. Because of the complex detailed features of the actual topography, it becomes necessary to make some simplifying approximations. We thus restrict our attention to that portion of the Reykjanes Ridge between 56° and 64° N latitude and in addition, take as the ocean floor the 2000 m depth contour; i.e., all depths outside the 2000 m contour are set equal to 2000 meters. We then approximate the topo-

Table 2

Laboratory Model

$D = 11$ cm
$H = 1.9$ cm
$v = 0.15$ cm/sec
$U = 0.15$ cm/sec
$\omega = 1.68$ rad/sec
$E = 2.1(10)^{-5}$
$Ro = 4.0(10)^{-3}$

graphy as a plunging ridge whose cross-sections are isosceles triangles. A sketch of the ridge is given in Figure 11. The vertical coordinate in the laboratory topography is exaggerated by a factor of 38.

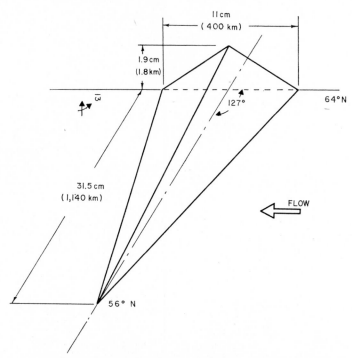

Figure 11 Model of a plunging ridge showing geometry used. Dimensions in parentheses are those for the Reykjanes Ridge

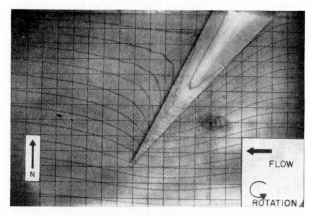

Figure 12 Mid-plane or free surface streamlines for flow over a plunging ridge (parameters are given in Table 2)

The laboratory free stream flow is taken parallel to "lines of constant latitude". Figure 11 is a photograph of the horizontal streamline pattern in the midplane of the tunnel which corresponds to the free surface streamlines. The prototype and laboratory parameters for this experiment are given in Table 2. Other experiments in the Ekman and Rossby number ranges of $2.0(10)^{-5} < E < 8.0(10)^{-5}$ and $0.002 < Ro < 0.09$ were also conducted. The flow patterns were however very similar to those shown in Figure 12. The photograph shows the strong topographic effect of the ridge. The streamlines are deflected to the southwest on the eastern side of the ridge and to the northeast on the western side. These results are in qualitative agreement with the current pattern sometimes observed over the Reykjanes Ridge. Note that the flow pattern over a plunging ridge is clearly different from that over a ridge of constant cross-section shown in Figure 9. Thus the slope of the topography along the ridge axis, although small, has an important effect on the flow characteristics.

IV DISCUSSION

The pattern of flow in the ocean and the model are in accord with the theoretical analysis of flow over topography by Welander (1969). To a good approximation the streamlines follow lines of equal f/h over the ridge where f is the Coriolis parameter, $2\Omega \sin \phi$, and h is the water depth. In the laboratory model we assumed a surface current that flows from east toward the west. The flow over the ridge deduced by Dietrich in June 1955 was clearly from south to north; however, the distribution of near surface water characteristics in June 1955 suggest that westward flowing currents prevailed for some time prior to his survey. Examination of Dietrich's maps of the salinity distribution at the surface, 200 m and 500 m, shows a broad tongue of saline water that extends from the eastern boundary of the Northeast Atlantic Basin toward the west between the 60th and 65th parallel. This distribution of isohalines suggests a general flow, that carries water with characteristics of North Atlantic water, westward from the area of the Faeroe Islands to the flanks of the Reykjanes Ridge. The flow pattern associated with dynamic topography observed by Dietrich probably was newly established and had not had time to destroy the distribution associated with the earlier westward flow. Thus there is evidence that previous to June 1955 the flow was in a direction similar to that assumed in our model.

These results further emphasize the large variations in location, direction and speed of flow of surface currents in the North Atlantic. These changes of flow could be an important factor in the volume of overflow water spilling over the Faeroe–Iceland and the Iceland–Greenland Ridges. The water behind each of these ridges is restrained by the dynamic "damming" effect of currents flowing along the southern flanks of these ridges. A decrease in velocity of these currents allows a greater discharge of overflow water into the North Atlantic basins.

Variations in the volume transport of the Norwegian Sea overflow water have been reported by several workers. Recently Talwani *et al.* (1970) reported evidence for an annual cycle in bottom water temperature over the flanks of the ridge between 1200 m and 1500 m. The evidence for these temperature variations are thermal waves detected in temperature versus depth profiles down to 12 m in the sea floor sediment. The above authors suggested that this variation was due to vertical oscillations of the upper boundary of the Norwegian overflow water, which in turn is related to changes in volume transport. The sediment temperature measurements further indicated that the bottom water temperature was near its minimum value in August 1966. Minimum temperatures occur when the overflow water upper boundary is at its highest level or volume transport and current velocity in the overflow water are at a maximum. As other workers have earlier suggested and our study further illustrates, there is a complex interaction between the deep flowing surface currents of the Northeastern Atlantic and the release of bottom water from the Norwegian Sea.

Flow Near the Floor of the Ridge

Further experiments were conducted in order to examine the flow characteristics of the boundary layers on the ridge surface. Figure 13a is a photo-

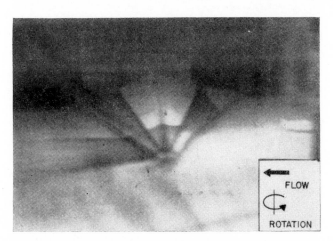

Figure 13a Flow along the surface of a plunging ridge in the Ekman boundary layer. The pattern of flow is shown by dye tracers released from two straight lines on the ridge surface. View is from above and facing northeast along the ridge axis

graph of one of these experiments. The view is from above and along the ridge axis from the southwest. The streaked surfaces observed are produced by releasing a dye tracer, using a technique introduced by Baker (1966), along two straight lines on the ridge surface. Figure 13b is an interpretive sketch of the photograph. The boundary layer transports are characteristic of Ekman layers; that is, the transport is to the left of the streamlines of interior flow, facing downstream (Figure 12). For the model then the Ekman layers tend to transport fluid away from the ridge axis on both sides of the

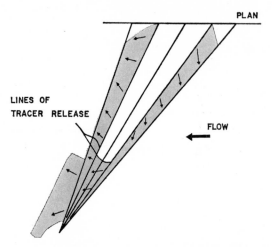

Figure 13b   An interpretive sketch of 13a in plan view showing the directions of boundary layer flow

ridge. The resulting flow away from the ridge axis along the floor of the ridge has an important effect on sedimentation over the ridge. There is strong evidence in the profiles shown in Figure 5 that bottom currents are influencing the distribution of sediment thickness.

First, it should be noted that the formation of a mid-oceanic ridge by spreading of the sea floor also has an effect on the sediment distribution. The sea floor on the Reykjanes Ridge is rifting apart at a rate of about 2 cm/yr (see for example Talwani *et al.*, 1970). Thus the basement rocks become progressively older with distance from the axis. At a distance of 100 km for example, the basement is about 10^7 years old. A uniform sediment accumulation on the sea floor would lead to a regular increase of sediment thickness with distance from the axis.

It is clear, however, that the sediment thickness does not increase in a regular way. For example at a distance of 100 km from the axis, the basement has little or no sediment whereas just beyond, the sediment becomes very thick indeed. Johnson and Schneider (1969) have surveyed this thick sediment body that appears as a long topography ridge and named it the Gardar Ridge. They attributed the formation of this ridge to sedimentation controlled by the Norwegian Sea overflow water. This sudden increase in the amount of sediment is due in part to a slowdown in spreading rate in the Atlantic between 10^7 and 4×10^7 years (Talwani *et al.*, in press, and Vogt *et al.*, 1970). During this period a much greater span of time was available for sediments to accumulate. However, the change in rate of spreading does not explain the dearth of sediment on the gentle scarp at 100 km from the axis and the V-shaped channel that is adjacent to the scarp on the eastern side of the ridge. These features, we think, result from the flow of overflow water along the contours. Note that the sediment is moved flankward as the Ekman transport experiments predict.

Such current controlled deposits are found in most areas of the sea floor where strong bottom boundary flow exists. In the northern hemisphere erosion occurs along the boundary to the right as one looks downstream, and mounded drift deposits are found to the left (LePichon *et al.*, in press), and vice versa in the southern hemisphere. The flow in the Ekman boundary layer may be an important mechanism in the formation of these drift deposits.

Acknowledgements

The geophysical survey was carried out under contract NONR 266(79) and N0014-67-A-0108-004 from the Office of Naval Research, and grants GA 1415, GA 1434, GP 5392, GA 5536, GA 1412, GA 18765, GA 550, and GA 1615 from the National Science Foundation. The model studies at the University of Delaware are supported by grant GA 943 from the Atmospheric Sciences Section of the National Science Foundation.

Manik Talwani was chief scientist on board the *Vema* when the geophysical survey was made. He also developed the computer programs that rapidly analyze the satellite navigation data. The efforts of H. C. Kohler, Captain of the *Vema*, and members of the ship and scientific crew are gratefully acknowledged. The assistance of Earl Gould in performing the model experiments is appreciated.

Discussions with Arnold L. Gordon (who was also responsible for bringing the two authors together) were valuable. Val Worthington kindly made the original data from the *Erika Dan* available.

References

Baker, D.J., A technique for the precise measurement for small fluid velocities, *Journal of Fluid Mechanics*, Vol.25, pp. 573, 1966.

Boyer, D., Rotating flow over long shallow ridges, *Geophysical Fluid Dynamics* (in press), 1970.

Boyer, D. and J.Guala, A model of the Antarctic Circumpolar Current in the vicinity of the Macquarie Ridge, *Antarctic Oceanology*, Vol. 2 (in press) 1972.

Cooper, L.H.N., Deep water movements in the North Atlantic as a link between climatic changes around Iceland and biological productivity of the English Channel and Celtic Sea, *Journal of Marine Research*, **14**, pp. 347–362, 1955.

Dietrich, G., Schichtung und Zirkulation der Irminger-See im Juni 1955, Ber. Dtsch. Wiss. Komm. Meeresforsch., 14, **4**, pp. 255–312, 1957.

Ewing, J. and R.Zaunere, Seismic profiling with a pneumatic sound source, *J. Geophys. Res.*, **69**, 22, pp. 4913–4915, 1964.

Hermann, F., Sections Faeroes to East Greenland and Cape Farewell to West Ireland, *Annales Biologiques* **11**, pp. 20, 1954.

Johnson, G.L. and E.D.Schneider, Depositional ridges in the North Atlantic, *Earth and Planetary Science Letters*, **6**, 416–422, 1969.

Lee, A.J., Temperature and salinity distribution as shown by sections normal to the Iceland–Faeroe Ridge, Rapports et Procès-Verbaux des Réunions, 157, pp. 100–137, 1967.

LePichon, X., S.Eittrein, and J.Ewing, A Sedimentary Channel Along Gibbs Fracture Zone, *J. Geophys. Res.*, **76**, pp. 2891–2896, 1971.

Pitman, W.C. III, and M.Talwani, Sea Floor Spreading in the North Atlantic (in press).

Talwani, M., C.Windisch, and M.G.Langseth, Jr., Reykjanes Ridge Crest: A detailed geophysical study, *J. Geophys. Res.*, **76**, pp. 473–517, 1971.

Vogt, P.R., N.A.Ostenso, and G.L.Johnson, Magnetic and bathymetric data bearing on sea floor spreading north of Ireland, *J. Geophys. Res.*, **75**, pp. 903–920, 1970.

Welander, P., Effects of planetary topography from the deep sea circulation, *Journal of Deep Sea Research*, supplement to Vol.16, pp. 369–391, 1969.

Worthington, L.V. and G.H.Volkmann, The volume transport of the Norwegian Sea overflow water in the North Atlantic, *Deep-Sea Res.*, **12**, pp. 667–676, 1965.

On the Structure of Deep Currents

V.G. KORT

Institute of Oceanography, Academy of Sciences
Moscow, U.S.S.R.

Abstract Direct observations of deep ocean currents in the western tropical Atlantic Ocean from the R/V *"Akademik Kutchatov"* are in agreement with the basic deep current structure found by the geostrophic calculations of Wüst (1957). Though the direct measurements indicate higher velocities, the agreement is remarkable considering the different methods employed and the 40 year time span between the observations.

In the well known paper of Georg Wüst (1957) on the transport of the water masses by the deep currents in the Atlantic Ocean Wüst shows for the first time the existence of the complicated layered structure of the deep ocean currents. However Wüst's conclusions on the layered structure of the deep current in the ocean aroused certain doubts, since they were based on direct methods of the analysis of the deep-water observations of the distribution of temperature, salinity and disolved oxygen but not on the instrumental current measurements.

Lately the development of technical means for such deep-water observations gives oceanographers the necessary data, though rather scanty, to study, by direct measurements of the deep oceans currents, the abyssal circulation. These data confirm the concepts of Wüst and have shown the complicated nature of the deep currents in the area of the western boundaries of the ocean.

From February to May, 1969, the scientific expedition of the Institute of Oceanology of the Academy of Sciences of the USSR on board of R/V *"Akademik Kurchatov"* and *"Dmitry Mendeleev"* studied the dynamics of the boundary current system in the western and tropical regions of the Atlantic Ocean (the Antilles, Guiana and Brazil currents) (Kort, 1969). Some regions of the instrumental current measurements coincide with those of the

"*Meteor*" Sections VI, VIII, XIII, XIV (1925–1927), the data of which were used by Wüst in his 1957 paper.

The comparison of Wüst's computations of deep-ocean currents with the direct measurements made more than 40 years later is very interesting both methodologically and from the point of view of the studies of the many-year water circulation measurements in the Atlantic Ocean. The meridional components of the geostrophic current from the "*Meteor*" data (Section VI) is shown in Figure 1a. Figure 1b shows section 7 through the Brazil current at 17° S latitude made by the "*Akademik Kurchatov*". The isolines of the velocities on this section correspond to the meridional component of the

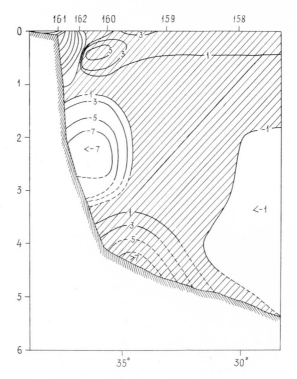

Figure 1a   Geostrophic velocities (cm/sec) for "*Meteor*" section VI (17° S), (Wüst 1957). Hatched area indicates flow to the north

average daily current velocity determined by the instrumental method. The littoral station 406 of the section is 40 miles of the continental shelf margin.

The comparison of these two sections show a wonderful similarity in the structure of the streams. Both the "*Meteor*" and "*Kurchatov*" data reveal the layered character (3–4 layers) of the current near the continental slope of Brazil. The identical major streams of the same direction are clearly seen on both sections. There are two opposite flowing streams in the upper 500 m layer: the littoral southward stream corresponding to the Brazil Current and northward stream east of station 403. A powerful flow of the North Atlantic

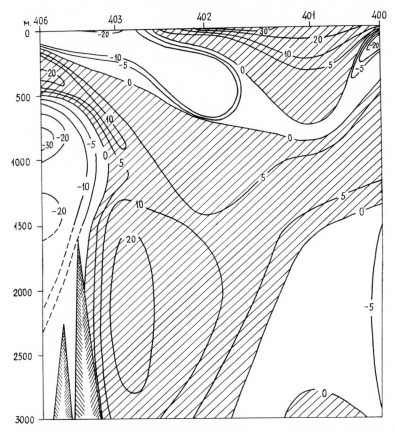

Figure 1b Directly measured velocities (meridional component in cm/sec) for the section 7 (17° S) of "*Akademik Kurchatov*". Hatched area indicates flow to the north

deep water moves to the south within the 500–1000 m to 3000 m layer near the continental slope. A northward bottom current joins the base of this stream on its eastern side. The greater part of the area of both sections is occupied by northward flow. The sections under consideration differ essentially in the velocity magnitude which in axis of the currents on the "*Meteor*" sections is about 3 times lower than the value of the average instrumental measurements. This difference is quite understandable since Wüst's data are related to the pure geostrophic transport, computed from the hydrological data of the stations 150 miles and more apart, i.e. from the considerably smoothed gradients of specific volume.

The *"Meteor"* Section XIII crossing the Guiana Current is shown in Figure 2a, and the *"Akademik Kurchatov"* Section 4 (approximately the same position as the *"Meteor"* section) is shown in Figure 2b. On the first section the meridional components of the geostrophic current velocity are

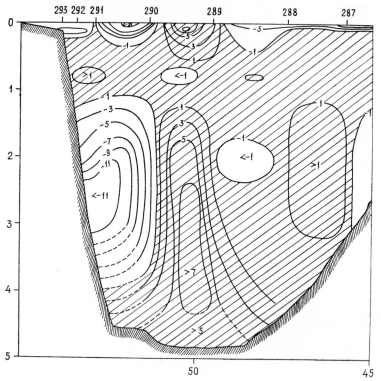

Figure 2a   Geostrophic velocities (cm/sec) for *"Meteor"* section XIII (Wüst 1957).
Hatched area indicates flow to the north

shown; on the second section the average daily current velocity component normal to the section is shown. In spite of some difference in the direction of the sections the structure of their deep currents is rather similar. Four streams of different directions are well pronounced in the surface layer on both sections. A rather strong southeast current flows on the oceanward side of the stream corresponding to the Guiana Current. There is a stream with the north component and still farther oceanward a stream with the south component. In both cases a strong southeast flow is found in the deep layers (1000–3000 m) flowing along the continental slope (Station 372 of the *"Kurchatov"* is situated thirty miles from the continental slope). The greatest velocities of the above-mentioned streams on the Section XIII as in the first case are much less than the averages computed from the *"Kurchatov's"* data.

The similar comparison of the measurement of the deep current of the *"Akademik Kurchatov"* data with the data of Sections VIII, XII and XIV of the *"Meteor"* shows the same similarity of the deep current structure in the western tropical Atlantic. The obtained result is particularly unexpected because the data on the deep current characteristics have been obtained by different methods by the expeditions separated by more than 40 years. This fact indicates: 1—the many-layer structure of the deep current is rather stationary in the western boundary regions of the Atlantic Ocean, at least, within the tropical zone, and, 2—the "core" method developed by Wüst in combination with a modified (choosing of two or three reference levels) method of computing the geostrophic currents is very effective in the dynamic analysis of deep hydrological observations.

The instrumental measurements of deep current velocities from the *"Akademik Kurchatov"* are indicative of a highly intensive water exchange in the deep layers of the ocean. For instance, the total discharge of all the

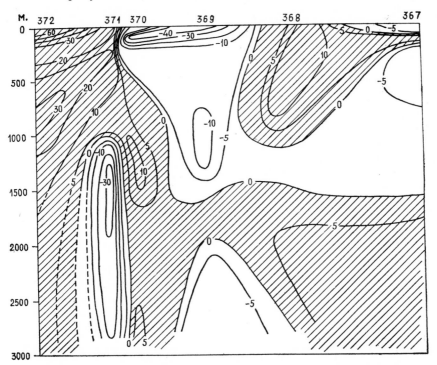

Figure 2b Directly measured velocities (meridional component in cm/sec) for the section 4, approximately same position of section shown in Figure 2a, of the *"Akademik Kurchatov"*. Hatched area indicates northward flow

northern streams on the Section 7 (Figure 1b) of the "*Kurchatov*" in a comparatively narrow coastal zone of the ocean (to 33° W) in the 0 to 3000 m
layer is 74×10^6 m^3/sec, whereas the geostrophic transport of waters northward through the whole Section VI of the "*Meteor*" crossing the Atlantic
Ocean along 17° S latitude, is according to Wüst 87×10^6 m^3/sec in the same
layer, and within the Brazil Basin -41.5×10^6 m^3/sec. The general water
transport in the southward streams of Section 7, including the Brazil Current,
is about 35 to 40×10^6 m^3/sec. For the whole "*Meteor*" Section VI the southward transport in the 0 to 3000 m layer is 99.2×10^6 m^3/sec and within
western basin -60×10^6 m^3/sec.

Thus, the modern direct measurements of the deep water transport show
a considerably more important role of deep water in the water exchange, in
particular, between northern and southern parts of the Atlantic Ocean. The
direct measurements of the deep currents show the existence of flow with
unexpectedly great velocities. The average daily current velocities of 15 to
20 cm/sec were recorded at 3000 m depth and of 10 cm/sec at 4000 m depth
almost on all the sections of "*Akademik Kurchatov*". The greatest values
of the instantaneous current velocities at the above-mentioned depths are
40 cm/sec and 20 cm/sec, respectively.

The isopycnic analysis of the deep hydrological observations from the
sections of the "*Akademik Kurchatov*" with the distances between the
stations no more than 30 miles show that the baroclinic conditions are found
to a depth of 3000 m and more in the region under study. This fact throws
light on the nature of the above-mentioned deep streams in the western
tropical Atlantic.

No less interesting facts were obtained lately for the western zone of the
tropical Pacific Ocean. On the 38th cruise of the R/V "*Vityaz*" (Kort, 1966)
it was established that 4–5 zonal streams of different directions exist in the
0 to 1000 m layer of the Equator between 132° and 151° E longitudes. In the
surface layer the current flows eastward; it is underlain by a layer where the
current flows westward, below the latter there is an eastward current and so
on. On the background of such a complicated structure is the equatorial
undercurrent (Cromwell Current), with the velocity of about 40 cm/sec in
the layer approximately between 150 and 3000 m. The region investigated
by the "*Vityaz*" includes the region believed to be the zone of generation of
the Cromwell Current (132°–135° E). The observations of this "*Vityaz*"
expedition show that a rather strong deep countercurrent exists under the
Mindanao Current traced to depths 1000–1200 m. The occurrence of well-
rounded large pebbles on the bottom of the strait between the Mindanao
Island and the Molucca Islands is evidence of the existence of bottom currents
about 2000 m with very great velocities in this region.

The numerous observations of deep currents in the Gulf Stream area carried out by the American and English oceanographers during the last ten years are well known. In the detailed review of these studies J. A. Knauss (1968) gives many original data about powerful deep countercurrents with average velocities of 10 to 20 cm/sec at depths of more than 2000 in the western North Atlantic.

Thus even these comparatively scanty measurements of the deep currents made to date show clearly enough an extremely complicated dynamic structure of the deep layers of the ocean. The existing ideas about extremely slow motion of deep water and about rapid attenuation of baroclinic and barotropic effects with depth evidently require serious re-examination.

References

Knauss, J. A., Water transport by the Gulf Stream. The Second International Oceanographic Congress. "Nauka" Publishers. Moscow, 1968.

Kort, V. G., 38th cruise of the R/V "Vityaz" (basic scientific results). Okeanologia, Vol. VI, No. 6, 1966.

Kort, V. G., Basic scientific results of the expedition on the R/V *"Akademik Kurchatov"* (5th cruise). Okeanologia, Vol. IX, No. 5, 1969.

Wüst, D., Stromgeschwindigkeiten und Strommengen in den Tiefen des Atlantischen Ozeans. Meteor-Werk, Band VI, Teil 2, pp. 261–420, 1957.

Suspended Particulate Matter in the Deep Waters of the North American Basin*

STEPHEN EITTREIM AND MAURICE EWING

Lamont-Doherty Geological Observatory of Columbia University
Palisades, N. Y. 10964

Abstract The turbulence associated with the vigorous bottom currents of the western North Atlantic maintains a nepheloid layer with an average thickness of one kilometer. The layer covers the continental rise, abyssal plains and Bermuda Rise. On the average, a tenfold increase in light scattering occurs from the clearest water above to the most nepheloid water near the bottom. Filtration of water samples taken in and above the nepheloid layer and analysis by microscopic particle counting establish an empirical relation between light scattering and particulate matter concentration. The light scattering is caused by suspended particles (and aggregates of particles) of (a) clays and silts, (b) silicate and carbonate test fragments and (c) other organic material. The concentrations in the most intense light scattering regions are about 0.1 ppm. The mean particle size (by volume) of particles, including aggregates, in the nepheloid layer is 12 μ. On the basis of the concentration gradients, and an assumed balance between downward gravitational settling and upward eddy diffusion, coefficients of vertical eddy diffusion are derived using Stokes' Law settling velocities. The values derived are on the order of 10^2 cm^2/sec. Such high intensity turbulence is facilitated by the extremely weak stability of the water column in the bottom one kilometer. Injection of clays and silts into the bottom waters by turbidity currents at the continental margin is believed to be the main source of the nepheloid layer particles. Such turbidity currents may be caused by slumping of unstable slope sediments or by tidal action on the shelf and slope.

INTRODUCTION

The knowledge about processes of sedimentation in the deep sea has rapidly expanded over the last few decades giving rise to at least two basic classifications of sediment: turbidite and pelagic. Interest in suspended sediments has increased recently because of the discovery of a class of sediments seen

---

* Lamont-Doherty Geological Observatory Contribution No. 1602.

on the seismic profiler which is neither turbidite nor pelagic but probably the result of a steady near-bottom flow process (Ewing and Ewing, 1964). Also, clean silt and sand laminae in sediment cores taken in deep sea lutite areas point to bottom current influence on sedimentation (Heezen and Hollister, 1964). Many bottom photographs taken in deep basins of the oceans, especially the Western Atlantic and Circum-Antarctic, where vigorous bottom water circulation exists, give very poor images of the bottom. This has been interpreted by some as caused by murky water. This evidence, together with the unanswered questions about transportation of lutites in the deep sea, has led to the gathering of deep-sea nephelometer data as a routine procedure on Lamont-Doherty ships, a program which has continued since 1965. The nephelometer used is essentially a modified Lamont-Doherty shutterless bottom camera with the addition of a continuous light source, a baffle-attenuator arrangement and a continuous film drive. This instrument was designed for simplicity, ruggedness and reliable operation at all depths on a day to day basis by relatively inexperienced personnel.

Quantitative measurements of light absorption and scattering require fairly sophisticated apparatus which is difficult to both maintain in its absolute calibration and to construct so as to withstand the rough treatment received by instruments lowered day after day to kilometer depths. Hence, the optical properties of sea water are among the least known physical parameters of the oceans. From the relatively limited number of measurements made in the last few decades by Kalle, Jerlov, Kullenberg and others, it has been shown that such measurements can be used in identifying certain water masses over limited areas, and in detecting large biologic populations. Most of this work has been summarized by Jerlov (1968).

This study is confined to the Lamont-Doherty nephelometer measurements made in the Northwest Atlantic Basin, with particular emphasis on the near bottom nepheloid layer. Ewing and Thorndike (1965) and Eittreim *et al.* (1969) discussed the measurements of this layer along the continental rise. The purpose of this paper is to extend this discussion to the whole basin. The nepheloid layer is apparently the agent of transport of most of the continental rise and probably the Bermuda Rise lutite deposits, and therefore is of great importance regarding the sedimentation in this basin.

INSTRUMENTATION AND DATA REDUCTION

A description of the nephelometer was given by Thorndike and Ewing (1967). The instrument consists basically of light source, camera and intervening baffle-attenuator. The baffle prevents direct light of the light source from entering the camera except for a small portion of it which passes through two calibrated attenuators. The direct light passing through the attenuators is used to monitor fluctuations due to changes in the intensity of the light source or in speed of film transport. The camera employs 35 mm film which is wound continuously past a 3/16″ slit. Minute marks are recorded on the film in order to correlate data on the film with length of wire paid out from the winch.

The nephelometer, which is usually mounted in the same frame as the bottom camera, is lowered continuously to the bottom by the hydrographic winch, thus providing a continuous record of light scattering during lowering and raising. After mid-1967 the nephelometers were equipped with bourdon tube-pressure gauges which recorded depth directly by the lateral deflection of a light spot along the margin of the film. Most of the data presented here were taken with nephelometers equipped with such pressure gauges, the exceptions being noted in Table 1. For data taken previous to 1967, depth is obtained by correlation of the log of wire out versus time with minute marks on the film.

The rate of film transport is approximately one inch per minute and lowering and raising rates are typically 100 meters per minute. With a slit width of 3/16″, a given spot on the film spends about 0.2 minute exposed to the slit. Therefore, the photographic arrangement acts as a running filter of width approximately 20 m, placing a limit on the thickness of water layer that can be resolved.

The intensity of scattered light determines the exposure of the film. To establish a scale of relative exposure, the camera is placed in a sensitometer prior to a lowering where 12 calibrated steps of increasing exposure (intensity) are produced on the film. The $\log_{10}$ ratio between exposures of adjacent steps is 0.2.

At a given distance out from the center of the film the exposure produced is the sum of scattering over a range of angles. The mean scattering angle is about 18° and about 90 % of the scattering is from angles between 7° and 30°. The minimum possible scattering angle allowed by the size of the baffle-attenuator is about 7°. Ambient light is recorded as "noise" on the film when in the vicinity of the surface (0–200 m) during the day. Since the downward irradience is reduced by about three orders of magnitude at 200 m in clear ocean water (Jerlov, 1968), this depth is believed to be a safe cut-off and accordingly no data is presented here for depths less than 200 m.

The constraints in the design of the Lamont-Doherty nephelometer prohibit measurement of the volume scattering coefficient, $\beta(\theta)$, or total scattering function, b, (Jerlov, 1968) which require control of wavelengths, volumes of water, angles of scattering and which require a well calibrated instrument. The objective here is an estimate of the total quantity of suspended sediment,

Table 1 Nephelometer stations used in this study

Cruise	Nephelo-meter Number	Latitude N	Longitude W	Bottom depth (m)	pressure recorder used	$\int E/E_0 \, dD$
Conrad 11	1	30° 38.0′	64° 37.0′	4782		6668
	3	21° 25.0′	57° 38.0′	5221		1898
	189	20° 33.0′	72° 22.0′	4152	×	956
	190	22° 04.0′	70° 50.0′	5502	×	9945
	191	22° 13.0′	70° 40.0′	5397	×	8180
	192	22° 24.0′	70° 29.0′	5368	×	8130
	194	23° 28.0′	69° 14.0′	5386	×	4214
	195	23° 50.0′	68° 53.0′	5335	×	5718
	196	24° 18.0′	68° 26.0′	5750	×	2977
	197	23° 59.5′	69° 35.0′	5432	×	4914
	198	24° 57.0′	72° 09.0′	5518	×	4991
	199	24° 43.2′	73° 47.0′	5313	×	6221
	200	25° 45.5′	73° 29.0′	5266	×	6510
	201	26° 16.5′	74° 02.0′	4830	×	3083
	202	26° 38.5′	74° 38.0	4402	×	3182
	203	26° 51.5′	75° 13.0′	4620	×	4574
	204	27° 04.0′	75° 41.0′	4715	×	4230
	205	27° 37.0′	76° 17.0′	4986	×	—
	206	29° 08.7′	76° 12.5′	5007	×	6921
	207	29° 25.0′	75° 47.0′	4993	×	6791
	208	29° 33.5′	75° 19.5′	4710	×	5055
	209	30° 04.5′	73° 59.0′	4298		3101
	211	33° 52.0′	71° 38.0′	5244		9100
	212	35° 31.0′	71° 31.0′	4322		5252
	213	36° 06.7′	71° 19.0′	4276		1942
	214	38° 01.0′	70° 53.0′	3387		4537
	215	38° 19.5′	70° 59.0′	3122		1107
Vema 25	65	24° 23.4′	48° 59.4′	5258	×	2311
	66	24° 46.7′	50° 26.1′	5102	×	2203
	67	25° 19.8′	56° 18.8′	5387	×	2113
	69	31° 25.5′	64° 47.3′	4280	×	7323
	70	37° 45.0′	71° 25.0′	3453	×	4675
	73	34° 51.0′	68° 25.0′	5236	×	6488
	74	35° 49.3′	68° 38.3′	4843	×	7497
	75	37° 23.0′	70° 15.0′	4196	×	3735
	76	37° 42.0′	70° 49.0′	3863	×	4046
Vema 23	2	38° 27.0′	57° 46.0′	5207		6214
	5	41° 55.0′	61° 21.0′	4181		3664
	6	40° 33.0′	60° 11.0′	4982		6974
	7	39° 35.0′	57° 40.0′	5221		8289
	8	37° 49.0′	48° 48.0′	5362		9380

Table 1   (cont.)

Cruise	Nephelo-meter Number	Latitude N	Longitude W	Bottom depth (m)	pressure recorder used	$\int E/E_0 \, dD$
Vema 23	9	38° 16.0′	45° 18.0′	5208		5808
	12	45° 05.0′	42° 02.0′	4409		—
	78	27° 26.0′	61° 18.0′	5775		2220
	79	27° 30.0′	61° 21.5′	5797		2448
	80	29° 00.5′	64° 22.0′	5005		2303
	82	34° 24.0′	60° 46.5′	4802		5405
	83	36° 12.0′	65° 38.0′	4921		6348
	84	36° 29.5′	66° 11.5′	5011		6364
	86	37° 07.0′	69° 33.0′	4353		5628
Gibbs 68-1	12	37° 23.0′	69° 43.0′	4263	×	4263
	13	37° 29.0′	69° 41.0′	4217	×	4217
	14	37° 16.0′	69° 49.0′	4278		—
	16	38° 17.0′	69° 27.5′	3698		—
Vema 22[1]	2	33° 15.0′	73° 12.0′	5227		
	4	30° 19.0′	74° 35.0′	3381		
	7	29° 05.0′	73° 23.0′	4512		
	86	30° 23.0′	47° 14.0′	3851		
	87	33° 29.0′	54° 10.0′	5543		
	89	38° 07.0′	65° 17.0′	4914		

[1] No sensitometer used; reduced to optical density of film.

and this is obtained by the empirical relationship, for a given depth, between amount of particulate matter collected from water samples and the scattering (film exposure) recorded by a nephelometer on the same wire.

In the data presented here the scattering units are normalized to the clearest water in the water column. If the assumption is made that the scattering in this clearest water does not vary appreciably from place to place across the basin, then data from station to station may be compared. A remarkable feature of the water column in the western north and south Atlantic is the wide expanse of relatively clear water which always exists over the mid-depth range. This clear water corresponds to the Middle and Lower North Atlantic Deep Water in the South Atlantic and the Middle and Upper North Atlantic Deep Water in the North Atlantic. Filtration of water samples taken at the depth of this clearest water in the North Atlantic on cruises *Vema* 25 and *Conrad* 11 showed concentrations of particulate matter about 0.01 ppm. As will be shown later samples taken deeper, within the nepheloid layer, showed considerably higher concentrations. Real variations in the scattering intensity of this clearest water from station to station are probably so small that they would not shift the true origin of the curves by more than ± 0.2 on the log exposure scale. Further refinement of light scattering techniques and new deep water data might allow such adjustments to be made, however. At the present time this is the best and most objective method available for a valid comparison of light scattering curves from station to station and the method has the great advantage that it compensates for any changes in instrumental constants or other parameters which may have occurred between stations.

The steps involved in the data reduction are the following: the 35 mm exposed negative film is transported through a recording densitometer which measures the optical density of the film along a line at a set distance out from the centerline of the film. Time or depth marks along the film are also recorded. The optical densities of the sensitometer calibration steps are obtained in a similar manner. An "H-D" or "characteristic" curve of photographic contrast can then be constructed from these calibration steps, plotting optical densities versus relative log exposures. The slope of the straight line part of the H-D curve is conventionally denoted by γ. For these data, γ averaged about 2.0 ranging from extreme values of 1.3 to 2.3. Optical densities recorded on the scattering part of the film are then converted to relative log exposure by interpolation on the H-D curve. The relative log exposures, log E, are normalized as described above. The units are therefore log E/E_0 with E_0 representing the exposure in clearest water.

The above operations are done by digital computer after digitizing the optical density of scattering and sensitometer parts of the film and the depth information. The normalized log exposures for both ascent and descent are presented as a function of depth in meters corrected according to Matthews tables (Matthews, 1939). The absolute depth accuracy is judged to be better than 100 m (Eittreim, 1970). Six of the stations (*Vema* 22) presented here were taken without having sensitometer patches put on the film and for these the curves are in arbitrary units of optical density.

Since film exposure is the product of light intensity and time of exposure, it will be proportional to intensity of scattering providing that the film speed is kept constant so that all points on the film spend equal amount of time exposed at the slit. The units are therefore effectively the log of the ratio of intensity of scattering at a point in the water column to the intensity in the clearest water.

OBSERVATIONS

This study is based mainly on the 61 nephelometer stations in the North American Basin located as shown in Figure 1.

The curves of light scattering versus depth are presented in Figure 2

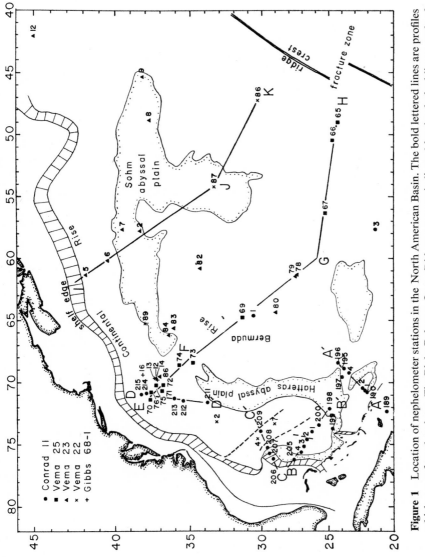

Figure 1 Location of nephelometer stations in the North American Basin. The bold lettered lines are profiles which are referred to later. The Blake Bahama Outer Ridges axes are indicated by the dashed lines south of Cape Hatteras.

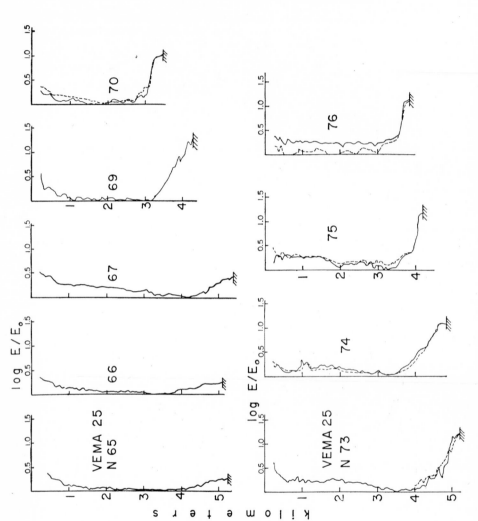

Figure 2 Profiles of light scattering versus depth for descent (solid line) and ascent (dashed line) of the instrument. Values of film exposure, E, are normalized to the exposure in the clearest water, E_0.

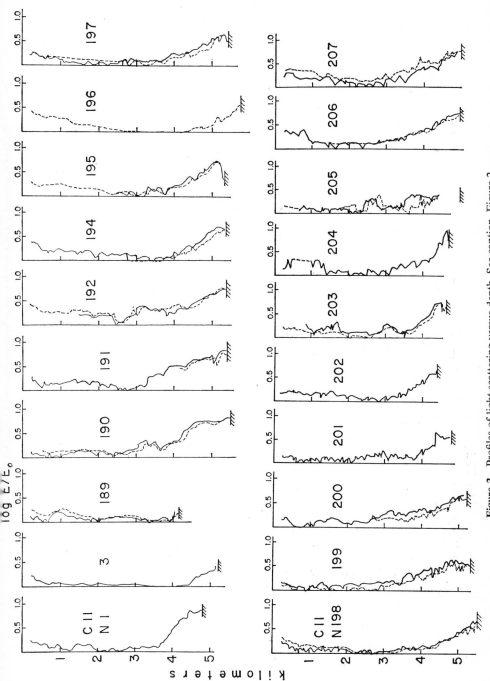

Figure 3 Profiles of light scattering versus depth. See caption, Figure 2

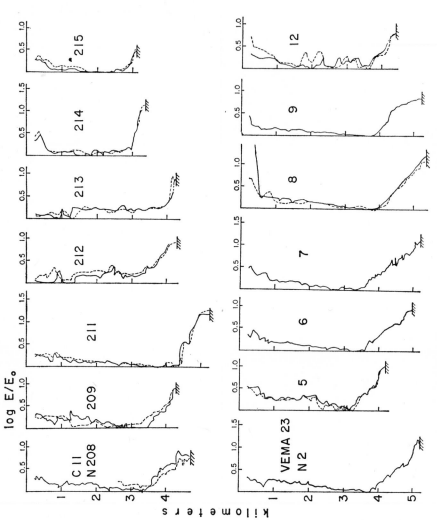

Figure 4 Profiles of light scattering versus depth. See caption, Figure 2

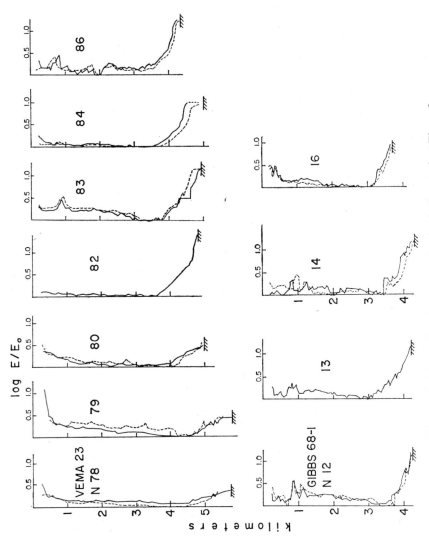

Figure 5 Profiles of light scattering versus depth. See caption, Figure 2

through 6. A typical curve shows a gradual decrease in scattering with depth reaching a minimum value at mid-water depths, then an increase in the bottom waters. The gradual decrease observed in the upper waters is probably a reflection of the downward decrease of organic particulate matter in the Northwest Atlantic (Gordon, 1970). Some of these data are contoured in basin-wide profiles E-F-G-H and I-J-K in Figure 7 and 8, respectively. The data along profiles A, B, C and D were discussed in Eittreim *et al.* (1969).

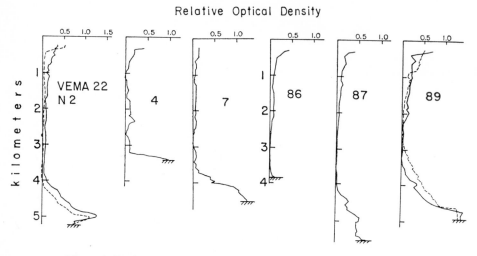

Figure 6   Profiles of light scattering versus depth for *Vema* 22 stations at which the films have no sensitometer control. Units are optical density of the film

The pronounced bottom nepheloid layer, which is the feature of major concern in this study, occurs at all of the stations west of the mid Atlantic ridge flank. It is most intense along the western boundary of the basin where sediment cover is thick and it is weak or absent over the abyssal hills region and lower ridge flank east of the Bermuda rise, where sediment cover is thin or absent as reflected by the rough topography on the eastern sides of both profiles. Note that stations V25-N65-67, between G and H, were taken in a fracture zone (Fox, *et al.*, 1969) and being deeper than the adjacent sea floor north and south, are probably not typical. See V22-N86 for what is probably a more typical ridge flank station. On the average, near-bottom intensities in the nepheloid layer are larger by a factor of 10 than the intensities in the clearest water at mid depths. It should be noted that the layer maintains its identity in thickness and intensity from the abyssal plain on the west all the way across the Bermuda rise.

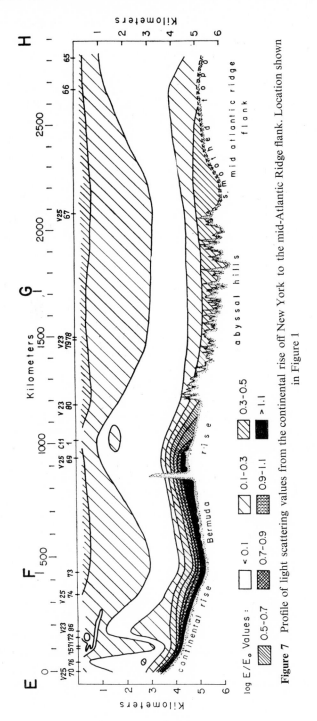

Figure 7 Profile of light scattering values from the continental rise off New York to the mid-Atlantic Ridge flank. Location shown in Figure 1

136

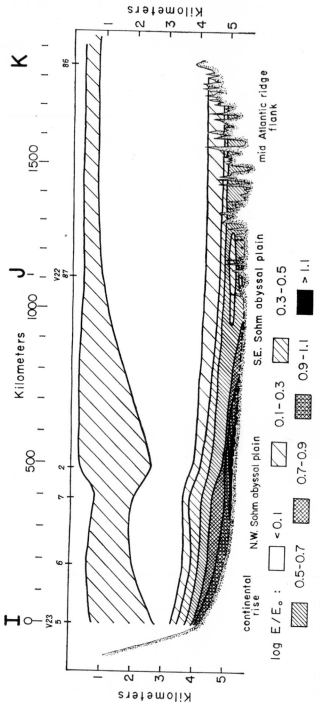

Figure 8 Profile of light scattering values from the continental rise off Nova Scotia to the flank of the mid-Atlantic Ridge. Location shown in Figure 1

To illustrate the variations of the nepheloid layer over the basin, the bottom part of the light scattering curves were drawn starting at the top of the nepheloid layer, the areas under the curves were blacked in and the curves located in their proper position on the map in Figure 9. The data show good correlation from station to station with similarly shaped curves ocurring over limited areas.

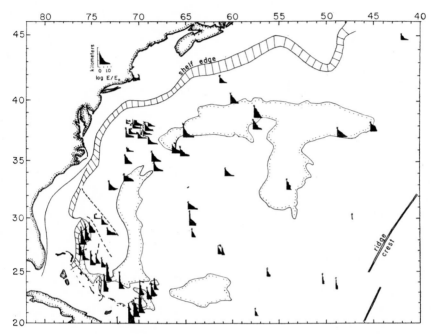

Figure 9　The lower portions of each light scattering curve comprising the nepheloid layer have been placed in their proper locations and the areas under the curves blacked in. The curves are the average between ascent and descent data. The actual locations of the stations are shown as x's

In certain instances, two or more stations which were taken in close proximity to one another afford an idea of the constancy of the layer over short distances, e.g. V23-N78 and 79 (Figure 5); G68-1-N12, 13 and 14 (see below and Figure 10) and possibly the cluster of about seven stations southeast of New York (Figure 9). The agreement among the three Gibbs stations is illustrated in Figure 10. These three stations were taken within thirteen miles of each other (table I) and the time lapses between them were 12 hours and 10 hours respectively. The "top" of the nepheloid layer (the first major departure from clear water) shows variation among the three from a depth of 3000 to 3500 m. However, all three curves have the same general shape, with a high gradient a few hundred meters above the sea floor and the same intensity of scattering at the bottom.

The concentration of particulate matter at any given depth should be proportional to the scattering intensity or film exposure recorded at that depth.

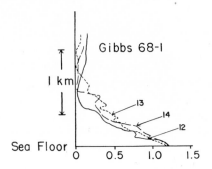

Figure 10   Data taken at three stations within 13 miles of each other on the lower continental rise off New York. Time lapses between stations were 12 and 10 hours respectively. The vertical axis is height above the bottom

This proportionality may not always hold, owing to variations in particle size, shape and index of refraction of the particulate matter involved in the light scattering process, but it is our working hypothesis. How well this hypothesis holds on the basis of particulate matter sampled in this basin and elsewhere will be discussed later. On the assumption of this hypothesis, if the exposure values are integrated over depth, a number will be obtained which will represent the total quantity of particulate matter in the nepheloid water column at that station. This computation was made for all the stations in the North American Basin whose curves were in units of relative exposure (all but V22), integrating from the top of the nepheloid layer to the bottom. The areas resulting were plotted on a map and contoured in order to illustrate the major trends (Figure 11).

A maximum follows the region of the lower continental rise—abyssal plain boundary as far south as the northern end of the Blake-Bahama Outer Ridge. The relative low over the Blake-Bahama Outer Ridge is at least partially a reflection of the fact that three of the stations in this area were on the ridge, in depths considerably shallower than the adjacent deep basins and were thus above and out of much of the nepheloid layer. This is illustrated in profiles B and C in Eittreim *et al.* (1969). South of this, a maximum again occurs in the deep western basin adjacent to the southern Bahama Islands, where depths reach 5.5 km. A more subjective contouring of the data, influenced by the bathymetric trends, would favor continuing the high values of $\int E/E_0 \, dD$ around the southeastern tip of the Blake-Bahama Outer Ridge. A relative high extends over the northwest half of the Bermuda Rise. The western flank of the mid-Atlantic ridge and southwest corner of the Bermuda Rise exhibit low values.

On the basis of the stations along the continental margin from New York to the Bahamas, Eittreim *et al.* (1969) noted an increase in thickness of the nepheloid layer and a decrease in near-bottom scattering intensity going from north to south. This was attributed to a source of sediments from the submarine canyons of the slope and rise in the region between New York and Cape Hatteras. As these sediment-laden bottom waters moved to the south, upward diffusion of sediment occurs, giving the layer its greater thickness. Since the Gulf Stream virtually cuts off terrigenous sources to the deep basin south of Cape Hatteras, reflected in the absence of canyons here, there is no addition of new sediment at the bottom of the water column and thus the

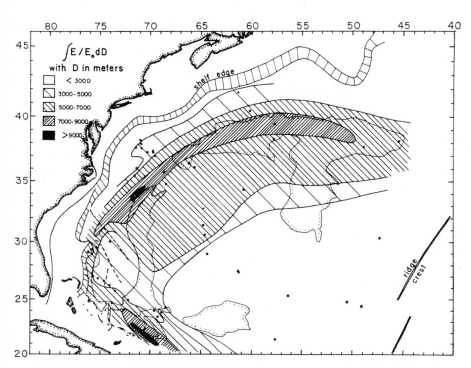

Figure 11 For each station the E/E_0 values have been integrated over depth, and the resulting values contoured as shown. Values of $\int E/E_0 \, dD$ are proportional to volume of suspended matter comprising the nepheloid layer at that station. Only those data which are in units of exposure were used. C11-N205 did not reach the bottom and was not used. Locations of the data are shown as dots. E/E_0 values used are the average between ascent and descent data

scattering intensity near the bottom weakens. This explanation requires a more or less steady supply of sediments from the canyons. Such a steady supply could come either from a steady-state turbid density flow down the axis of several canyons such as observed off the west coast of North America (Moore, 1966; Shepard *et al.*, 1969) or from individual episodic turbidity current flows at a rate of at least several per year. Considering the large total number of submarine canyons off the east coast, the rate of recurrence of turbidity flows in any one canyon in this case need be only on the order of a few per hundred years. These of course would be minor events compared to the far-reaching turbidity flows which produced the recognizable layers of graded bedding in the abyssal plains. Perhaps the lack of evidence in the sedimentary record of such minor and frequent flows argues for the previously mentioned steady kind of turbid down canyon flow. Erosion from the bottom as a sediment source to the nepheloid layer could not be ruled out, although the lack of persistent strong erosion of the northern continental rise coupled with its large thickness of the recent sediments (Hollister, 1967) tends to argue against it. If such erosion is a major factor, it is likely to be more prominent north of Cape Hatteras where the near-bottom high intensity light scattering is found.

In July 1966 a series of closely spaced nephelometer stations were made from R.V.Trident by E.Schneider and P.J.Fox between New York and Bermuda. No sensitometer patches were put on the film so that it was not possible to convert to exposure units. Also the film was not developed uniformly. For these data, the variations in optical density of the direct trace were subtracted from scattering trace variations to partially correct for changes in developing along the film. This improved the agreement between ascending and descending curves considerably. The locations of these stations and a section with the curves superimposed on topography are given in Figure 12. The units are arbitrary with values of scattering taken as zero minimum to 10 maximum. No bottom light scattering layer or at most a very weak one is seen above about 2500 m on the upper rise. The profiles show approximately the same configuration of the nepheloid layer as shown in E-F or I-J (Figures 7 and 8). The extremely close station spacing gives a profile which is valuable in that it establishes, beyond doubt, the continuity of the nepheloid layer down the rise.

Nephelometer data taken on *Vema* 23 between the Flemish Cap and Greenland was reported by Jones *et al.* (1970). Nepheloid layers comprising the bottom 300 to 1000 m of the water column exist along the northern (adjacent to Greenland) and southwestern margin of the Labrador sea with relatively weak light scattering in the bottom waters of the central part. Scattering intensities in these layers are roughly comparable to those recorded

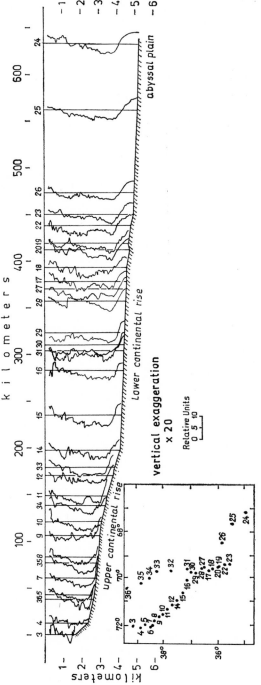

Figure 12 Locations and profile of data taken on R.V. *Trident* in July 1966. The light scattering units are arbitrary with zero for the minimum scattering to 10 for maximum. Profiles are the average between data for descent and ascent. Vertical lines give the location of the stations and are the mid point, or the unit of 5 on the arbitrary scale of light scattering

along the western side of the North American Basin. Approaching Cape Farewell on both south and east sides, there is an extremely sharp gradient in light scattering with very high values in the bottom 50 to 100 m.

BOTTOM WATER CIRCULATION, WATER MASSES AND DENSITY STRUCTURE OF THE DEEP WATERS

Wüst (1933, 1935) suggested a western boundary current in the North Atlantic from the observed high in the oxygen values in the core of the Lower North Atlantic Deep Water. From the theoretical work of Stommel (1957), measurements and dynamic computations of Swallow and Worthington (1961), Volkmann (1962) and Barrett (1965) and bottom photographic evidence (Heezen *et al.*, 1966), the existence of the Western Boundary Under-current (WBU) is firmly established. Velocities measured in it range from 8 to 18 cm/sec.

The water masses contributing to this current are difficult to define, but it is generally agreed that they consist of cooled Arctic and sub-Arctic waters which descend to the bottom around Greenland and Iceland. Wüst termed these waters Lower North Atlantic Deep Water (LNADW). Baffin Bay water is less dense than that from the other Arctic sources and hence, at present, is not thought to be a contributor to the WBU (Dietrich, 1961).

The other major bottom current in the Northwest Atlantic, the Antarctic bottom current, AABC, (which is a current responsible for the spreading of Antarctic Bottom Water, AABW) reaches as far north as 40° N according to Wüst. He concluded that the AABC branched upon entering the North American Basin, one branch flowing west of the Bermuda Rise, the other to the east of it. Since it would hug the right-hand side of any basins through which it traversed in the northern hemisphere, the western branch would circulate clockwise around the perimeter of the Bermuda Rise and the eastern branch would follow the base of the Mid-Atlantic Ridge flank.

Besides this general pattern, some mixing and recirculation of the two bottom water masses (Arctic and Antarctic) occurs. Amos *et al.* (1970) show that some Antarctic Bottom Water turns counter-clockwise around the margin of the Hatteras abyssal plain and is entrained in the southerly WBU flow at about 35° N. Volkmann (1962), to explain the extremely high trans-port values calculated for the WBU from data on the continental rise off Cape Cod, believed that some deep Gulf Stream water may be recirculated to the south in the WBU in the region north of Cape Hatteras.

Besides evidence by Heezen *et al.* (1966a) from bottom photographs showing the contour-following southerly flow of the WBU south of Cape Hatteras, Schneider *et al.* (1967), Heezen *et al.* (1966b) and Rowe and Menzies (1968) by similar evidence further support the pattern of southerly flow along the continental margin and northerly flow along the western Bermuda Rise. The circulation pattern on the Bermuda Rise is not well defined however as Knauss (1965) has measured southeasterly bottom currents on its western side, in conflict with the above evidence from bottom photographs.

A major source of the WBU was delineated by Stefansson (1968) with two hydrographic sections across the Denmark Strait. His temperature and salinity sections show a strong southward flow of cold bottom water across the Iceland–Greenland Ridge. Swallow and Worthington (1969), measured near-bottom velocities and transports across the Labrador Basin using neutrally buoyant floats and made geostrophic computations based on a reference level at 1200 m. According to their measurements, the flow is northwesterly southwest of Cape Farewell and southeasterly north of the Flemish Cap, consistent with a general counterclockwise geostrophic bottom circulation around the Labrador Sea. The bottom circulation in this area is further discussed by Jones *et al.* (1970) and related to the distribution and characteristics of the sediments of the basin.

From the data discussed above, Figure 13 has been prepared to summarize the known bottom water circulation in the northwest Atlantic. Comparison of this map to those defining the nepheloid layer (Figures 9 and 11 and Jones *et al.*, 1970) leaves little doubt that the two are related. Bottom circulation is strongest in this basin along the western margin of the North American Basin and around the perimeter of the Labrador Basin, places where the amount of suspended matter in the deep water is greatest. In general strong bottom nepheloid layers of the world are in areas where bottom water circulation is vigorous: circum Antarctic (Eittreim *et al.*, this volume), Argentine Basin (Ewing *et al.*, 1971), central western Atlantic (M. Ewing, pers. comm.), the far northeast Atlantic in regions of overflow water from the Arctic (Jones *et al.*, 1970), and the Northwest Atlantic. Areas where bottom water circulation is sluggish such as the Caribbean, Eastern Atlantic north of Walvis Ridge and south of European Basin (Connary, 1970), much of the Pacific Ocean (Ewing and Connary, 1970) and most of the Mediterranean (Ryan, 1970) have generally no bottom nepheloid layers or very weak ones. In the Arctic Ocean, Hunkins *et al.* (1969) find a good correlation between the occurrence of bottom nepheloid layers and bottom currents. They find that the deep, flat abyssal plain areas are devoid of both nepheloid waters and evidence of bottom currents, whereas the ridges around these deep basins have appreciable bottom currents and bottom nepheloid layers associated with them.

From Figures 11 and 13 it is seen that the vigorous WBU composed of

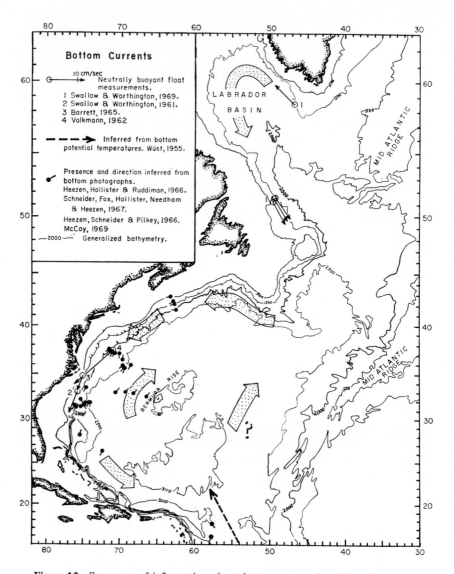

Figure 13 Summary of information about bottom water circulation from near-bottom current measurements and bottom photographs. Current measurements are from neutrally buoyant float data of Woods Hole workers. Only data which was taken closer than 1 km to the bottom is included. The arrows shown at locations 2 and 3 are vector averages of numerous measurements. The large arrows indicate idealized flow pattern derived from the measurements. See discussion in text

Lower North Atlantic Deep Water is identified with high turbidity while the Antarctic Bottom water which composes a significant component of the abyssal waters in the eastern side of the deep North American Basin is relatively non-turbid. In the Southwest Atlantic, the identity of these water masses in regard to their turbidity is reversed, with clear North Atlantic Deep Water overlying highly turbid Antarctic Bottom Water (Ewing *et al.*, 1971). Thus in general the LNADW decreases its particle content toward the south while the AABW decreases its particle content toward the north. Turbidity then, apparently is a relatively nonconservative property of water masses, being more closely related to the relative vigor of movement of the bottom water mass and perhaps proximity of detrital sources than to any process at its origin.

For the maintenance of a basin-wide nepheloid layer, turbulence is required to keep the particulate matter in suspension. High density stratification of the water column inhibits vertical turbulence, since the water particles must overcome buoyancy forces as well as friction to move vertically. Since salinity generally decreases downward in the bottom waters of the Northwest Atlantic (Fuglister, 1960), the temperature gradient is a good indicator of relative stability. The greater the temperature gradient, the more stable the water column.

A temperature section of Fuglister (1960) along 36° N is shown in Figure 14. The top of the nepheloid layer, as observed at stations nearest 36° N is superimposed on this temperature section. The important point to note is the lack of strong gradients in deep waters below the top of the nepheloid layer. In the listed data of Fuglister (1960) many adiabatic temperature increases near the bottom are observed along profiles 32° N and 36° N. Thus, the water of the nepheloid layer in the North American basin is relatively less stably stratified than the waters above. Thus the large scale turbulence which seems to be required by the quantity of material in suspension is not hard to accept for these bottom waters.

An illustration of the extremely weak density stratification in the deep waters in the North American Basin is shown by the potential density section of Lynn and Reid (1968) for the Northwest Atlantic. The isopleths become very widely spaced in the waters below about 3000 m. Hesselberg's (1918), criterion for static stability is:

$$E = \frac{1}{\varrho} \frac{\delta\varrho}{dz}$$

where $\delta\varrho$ is the density difference between a parcel of water which has been displaced from its usual depth (by a distance dz) and the surrounding water at this new depth.

10 Gordon II

The potential densities given by Lynn and Reid are the densities which the water would have if moved adiabatically to the 4000 m level. Therefore, values of $\delta\varrho$ and dz may be obtained from their contours of potential density by taking the contour interval and the distance between the contours respectively, at a location where one contour crosses the 4000 m level. Such a

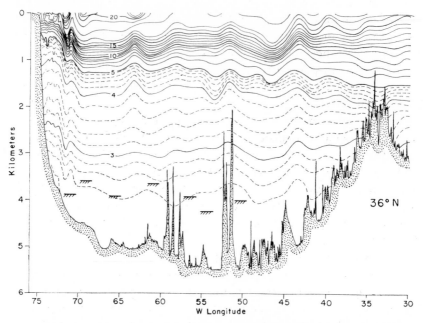

Figure 14 *In situ* temperature profile along 36° N. Depth of the "top" of the nepheloid layer, projected onto the section from the nearest nephelometer stations, are indicated by the hatchured lines. "Top" of the layer was taken to be the first departure from clear water. Nephelometer stations used, from left to right are C11-N213, V25-N74, V23-N84, and 83 averaged, V23-N82, 2, V22-N87, V23-N8. Modified after Fuglister (1960)

calculation gives $E = 0.05 \times 10^{-8}$ for the waters at the 4000 m level and even smaller values at greater depth. This value of E, is so small that it may not be significant. The density gradients at this depth are so small in fact that it has been suspected by some that they may be below the limit of resolution of density determination by the technique of computation from T and S data. Wüst (1933) suggested such an error to be the cause of what he calculated to be an unstable density stratification (E negative) in the deep waters here, a situation he could not accept as real. Lynn and Reid (1968), however, point out that potential densities, when referred to the 4000 dbar level maintain a stable, although extremely weak, stratification.

PARTICULATE MATTER CONCENTRATIONS

Because of the extremely low concentrations of particulate matter in the nepheloid layer, measuring these concentrations with precision is difficult. The data presented in this chapter were obtained by use of 2 liter "Niskin Sterile Samplers" (Niskin, 1962) lowered on the same hydrographic wire as the nephelometer. From three to five water samplers, arranged particularly to sample the transition from clear water to nepheloid water and the water of maximum turbidity near the bottom, were filtered immediately after collection on board ship using $0.45\,\mu$ pore size Millipore filters. The samplers employ a disposable polyethylene bag the opening to which is kept closed until the instant of collection.

Upon receiving the sample on board, the clamped neoprene tube from the bag is led directly to the filter holder top, the tube is unclamped and the water drawn through the filter with a vacuum pump. After filtering the entire sample, the filter is rinsed with 60 ml of distilled water. Contamination by particulate matter in the rinse water is minimized by prefiltering the rinse water through a $0.45\,\mu$ filter as it is used. By preventing any exposure of the sample to the atmosphere and by using a new polyethylene bag for each sample, contamination is kept to a minimum.

The filtered residues were analyzed by microscopic particle counting, using phase contrast viewing. Counts were made in eight size classes from $0.625–1.25\,\mu$ to $80–160\,\mu$. Preparation for viewing involved mounting a segment of the filter on a glass slide in oil of refractive index 1.51, which rendered the cellulose acetate filter material transparent. The phase contrast equipment allowed viewing of the fine clay particles which under ordinary or polarized light are transparent due to their small thickness. The highest magnification used was $\times 600$. The smallest particles capable of being detected are about $0.1\,\mu$ although none below $0.625\,\mu$ were measured and counted.

Particle size was defined by the intermediate diameter, which, if it is assumed that the smallest diameter lies normal to the filter surface, is the smaller of the two diameters observed. The particle diameters were measured using a micrometer ocular.

An average of 300 particles were counted for each sample. The total computed number, N, of particles on the filter in each size class was multiplied by the square and the cube of the mean diameter, d, giving numbers representing total particulate surface area, and total particulate volume, respectively for that class. These values of Nd^2 and Nd^3 were then summed up for all classes, giving total surface area and total volume. Absolute surface areas or volumes can be derived by the relationship between d^2 or d^3

and the expression for the true surface area or volume of whatever particle shape one assumes.

Jerlov (1955) experimentally substantiated the theoretically sound hypothesis that light scattering is proportional to total surface area of particles. He measured scattered intensities produced by various concentrations of particulate matter and showed that when the concentration of suspended matter is expressed in surface area of particles per volume of water (cm^2/L), the relationship between amount of particles and the scattering produced is linear. Beardsley *et al.* (1970) showed the linearity also to hold generally for surface water oceanic particulate matter. Jerlov (1955) also substantiated that a significant decrease in scattering occurs for particles smaller than 2μ a result produced by the decreased efficiency of the smaller particles to scatter light. Since light scattering is most closely correlated to surface area of the particulate matter, the value Nd^2, per volume of water collected, should be directly related to the scattering intensity which that water sample would produce.

Many sources of error arise in an analysis of this sort and one should be aware of the magnitude of the inaccuracies which they cause in the final results. They are, in the order of their importance:

1   Definition of particles: a range of particle types exists from clearly individual particles to loose aggregates of a number of particles. Aggregates or "flocculates", may consist entirely of clay and silt mineral particles but more commonly are a combination of clays-silts and organic material (Folger, 1968, frequently observed this in surface waters). These aggregates are judged to be either formed in the water, or only apparent aggregates formed by individual particles settling side by side on the filter during filtration. The former are counted as particles, the latter are not and to distinguish between them is sometimes difficult. Errors due to this lack of a clear definition of the category "particle" are probably less than 20% in the surface area determination, but may cause larger errors to be introduced into the relationship between surface area and light scattering for a given individual sample. Examples of these aggregates are shown in Eittreim (1970).

It should be pointed out that flocculation of the fine clays into larger particles with considerably higher settling velocities is probably a major mechanism by which the suspended matter reaches the bottom (Einstein and Krone, 1962; Whitehouse *et al.*, 1960) and therefore the importance of these aggregates should not be underestimated. In sedimentological analyses of deep sea lutites it is common practice to break down these aggregates into their individual particles by means of dispersing agents. After breakdown of the aggregates, mechanical analyses are done by separation of particles according to their Stokes settling velocities. This is, of course, the only practical objective method of making mechanical analyses of the fine clays, but if one takes the particle sizes derived in this manner literally, it can lead to false conclusions regarding the actual settling velocities in sea water.

2   Uneven distribution on the filters may cause errors. Although a random distribution of microscope fields is taken, the number of these fields is for practical reasons limited. Absolute values of particulate matter amounts were not used from the filters which exhibited obvious unevenness, but the data from these filters was considered still good for size-frequency analysis.

3   On the filters with the densest residue, where there is overlap of particles, many smaller particles may go uncounted.

4   Some samples contained a significant number of particles smaller than $0.625\,\mu$ the smallest size counted. This error would result in an underestimation.

5   Using the median diameter of the size class as a representative average for the particles in that class gives rise to error. This error decreases as the number of classes chosen increases, with no error for an infinite number of classes. For the largest two size classes, 40–$80\,\mu$ and 80–$160\,\mu$, where this error would be largest, particle sizes were actually measured and a true mean diameter was obtained, for each class. Since, in general, the true mean will be smaller than the median diameter between the end points of the class, this error will tend to cause an overestimate, in the opposite direction as numbers 3 and 4 above.

It is clear that with these uncertainties, one should not expect a simple straightforward relationship between light scattering and concentration sample-for-sample However, if enough samples are obtained, and a statistically valid relationship is seen between the two measurements, then that relationship can be used for the general purpose of deriving an estimate of one quantity from the other. It seems unlikely that any of the above errors (the most worrisome of which are numbers 3 and 4) could cause a persistent error in the end result larger than a factor of two.

Water samples, in conjunction with nephelometer lowerings were taken at 25 stations between the Bahamas and New York on *Conrad* 11, seven stations on the continental rise between New York and Bermuda on *Vema* 25 and 11 stations on *Gibbs* 68-1 between Bermuda and New York. Three samples were taken at each station on *Vema* 25 and at most stations on *Gibbs* 68-1. These were spaced, on the basis of previous nephelometer data, to take two samples in the nepheloid layer and one in the clearest water above. On *Conrad* 11, five samples were taken at each station, three in the nepheloid layer, one in the clearest water in mid depths and one in the surface waters at 100 m depth. The data from the particle count analysis of these samples are given in Table 2.

Table 2

Cruise	Nephelometer Number	Water Sample Number	Location N	Location W	Distance above bottom (m)	Volume collected (L)	Particle conc. Nd^2/vol. (mm²/L)	Mean size by area μ	Mean size by vol.² μ	Light scattering at depth of sample E/E_0
Vema 25	70	V1	37°45'	71°25'	549	2.70	3.32	3.5		1.7
		V2			146	—[1]		2.0	9.0	11.5
		V3			55			5.5	25.0	2.0
	72	V4	36°52'	70°14'	549	2.35	7.49	8.0		1.2
	73	V6	34°51'	66°25'	1280	2.48	7.82	7.0		
		V7			549	—[1]	2.74	4.0		
		V9			82	4.03	16.6	2.0	15.0	15.2
	74	V10	35°49'	68°38'	1280	—[1]	2.69	19.0		3.0
		V11			732	2.60	19.27	2.2	14.0	
		V12			73	1.92		2.0	17.0	20.0
	75	V13	37°23'	70°15'	1280	4.06	2.56	6.0		1.5
		V14			549	3.20	1.47	5.0		1.9
		V15			55	2.94	12.28	3.0	20.0	20.0
	76	V16	37°42'	70°49'	915	—[1]	3.62	2.3		2.4
		V17			366	2.30	12.76	2.2		
		V18			55	1.63		2.7	16.0	13.0
Conrad 11	190	4-1	22°04'	70°50'	37	—[1]		2.3	6.1	
		4-2			219	—[1]		2.4	11.5	
		4-3			549			2.4	8.5	
		4-4			2744	1.45	1.23	4.0		1.2
	191	5-4	22°13'	70°40'	219	1.85	5.50	2.4	14.0	5.0
		5-5			22	2.07	1.80	4.0	9.5	6.2
	192	6-1	22°24'	70°29'	22	—[1]		4.0	11.0	
		6-3			549	1.50	1.01	5.8	18.0	2.9
	194	8-1	23°28'	69°14'	22	—[1]		2.5		
		8-2			219	2.35	2.27	1.5	9.0	4.5
		8-3			549	—[1]		5.5	15.0	
		8-4			2744	0.77	2.34	11.0	12.0	
		9-1			22	—[1]		5.4	15.0	1.3
	195	9-2	23°50'	68°53'	219	—[1]		4.6	13.5	
		9-3			549	2.53	1.91	5.0	11.0	3.3

	Sample	Latitude	Longitude						
Conrad 11									
196	10-1	24° 18'	68° 26'	22	1.18	1.15	7.5	15.0	3.7
	10-2			219	1.56	2.64	6.0	12.5	2.8
	10-3			549	1.22	2.96	6.0		2.0
197	11-1	23° 59.5'	69° 35'	22	3.30	3.82	3.5	10.0	3.8
	11-2a			221	2.35	2.11	4.8	11.5	3.8
	11-3			549	1.98	1.77	4.3	11.0	2.5
198	12-1	24° 57'	72° 09'	22	4.13	3.22	2.0	7.0	4.5
	12-2			217	3.28	2.54	2.4	14.5	3.3
	12-2a			221	3.20	3.68	1.9	5.5	3.3
	12-3			549	2.47	3.41	4.8		2.4
209	21-1	30° 04.5'	73° 59'	37	1.75	2.40	3.5	4.7	7.9
	21-2			274	[1]		3.7	11.0	
211	24-1	33° 52'	71° 38'	37	[1]		9.5	18.0	
	24-2			274	[1]		2.1	6.5	
	24-3			549	[1]		3.5	11.0	
	24-4			1280	2.75	4.38	9.0		1.1
212	25-1	35° 31'	71° 31'	37	[1]		3.5	13.0	
	25-2			274	[1]		4.2	9.5	
213	26-1	36° 06.7'	71° 19'	37	[1]µ		3.5	12.0	
214	27-1	38° 01'	70° 53'	37	[1]µ		4.5	13.0	
	27-2			256	3.00	3.25	5.5	12.5	3.9
	27-3			457	[1]		6.0		
Gibbs 68-1									
4	G-9	34° 15'	66° 42'	476	[1]		3.2		
	G-10			55	[1]		6.0		
9	G-16	34° 27.5'	69° 34'	549	[1]		4.0		
	G-17			183	[1]		3.2		
10	G-18	35° 43'	69° 49'	2561	[1]		2.7		
	G-19			457	[1]		3.5		
	G-20			27	[1]		3.5		
13	G-24	37° 29'	69° 41'	110	[1]		3.0		
	G-25			311	[1]		3.1	9.0	
	G-26			27	[1]		2.2		
14	G-27	37° 16'	69° 49'	1006	[1]		3.0		
	G-28			512	2.45	3.04	4.2		
	G-29			476	2.50	2.02	6.5	6.0	1.9
	G-30			55	[1]		3.0		2.0

1 Uneven distribution on the segment of the filter analysed. These were judged unsuitable for making absolute determinations of particulate matter quantity.
2 Mean sizes by volume were determined for samples taken in the nepheloid layer only ($E/E_0 > 2.5$).

No detailed identification of particles was attempted. However, on the basis of particle shape, crystallinity and birefringence, some tentative identifications could be made. The materials believed to be present in roughly the order of abundance were:

1 Clay and silt sized mineral particles.
2 Amorphous organic "humus" or protoplasm, and other nonminerogenic parts of organisms.
3 Siliceous and calcareous organic tests, mostly diatoms and radiolaria, some foraminifera, coccoliths and sponge spicules.
4 Opaque irregularly shaped particles, probably industrial ash.
5 Pollen and spores.

Unfortunately, definite quantitative identification of such fine-grained particulate matter probably can only be made by x-ray diffraction, electron-probe and other indirect methods and requires larger amounts of material than obtained here. Although quantitative identification of particles could not be made, the clays and silts appear to be the predominant fraction in the nepheloid layer comprising anywhere from 60 to nearly 100% of the particles. In samples above the nepheloid layer, non-clay and silt particles were often predominant. Photomicrographs illustrating these differences are given in Eittreim (1970).

Size-frequency histograms and cumulative frequency curves were constructed for each sample and the median sizes are given in Table 2. As mentioned previously, the smallest size class taken, for practical optical reasons was 0.625–1.25 μ. For about half of the samples, a significant portion (greater than about 10%) of the residue was smaller than 0.625 μ. For 30% of the samples, the smallest size class contained the most particles, by surface area, of any class, indicating that for these samples about 20% to 60% of the particles were not counted.

Histograms of median sizes in and above the nepheloid layer are plotted in Figure 15. A significant difference in particle sizes is seen between samples in the nepheloid layer and samples taken above the layer. The explanation for these smaller sizes in the nepheloid layer is that above the nepheloid layer, organic debris comprises a larger fraction of the particulate material and this tends to be larger in size than the clay and silt particles which predominate in the nepheloid layer below. Much of this larger-sized organic particulate matter may dissolve in the deeper waters, but most likely the particle size difference is mainly due to the increase in the clay and silt particles in the deeper water.

No consistent variation in median size was observed with distance from the bottom in the nepheloid layer. The medians of the median particle sizes (diameter) for all samples in the nepheloid layer are 3.4 μ by surface area and 12 μ by volume (Figure 15). Therefore, it appears that in general a predominant portion of the suspended matter falls above the size of 2 μ, the size below which the extinction coefficient or scattering efficiency begins to decrease. However, these smaller particles are present and their optical effect undoubtedly introduces a further error in the scattering-concentration

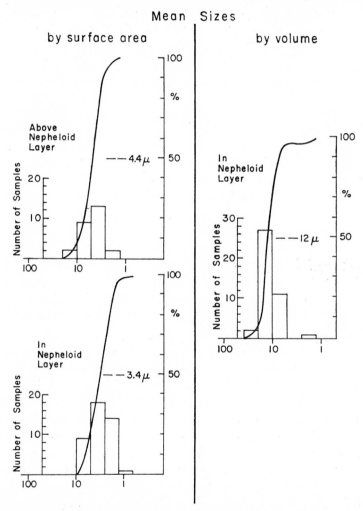

Figure 15 Mean sizes (intermediate particle diameter) of particulate matter samples above and in the nepheloid layer, by surface area and by volume. "In" the nepheloid layer is defined as at depths where $\log E/E_0 > 0.4$. Right hand abscissa refers to the cumulative frequency curves. Particle size is in microns and the medians of the median sizes are indicated. As mentioned in the text, aggregates are considered as single particles

relationship. From measurements in the surface waters in the Pacific, Beardsley *et al.* (1970) estimate that approximately $\frac{1}{2}$ of the recorded light scattering is due to particles smaller than 1 μ.

The main result of this water sampling program is given in Figure 16. Total amounts of suspended matter were determined for only those filter segments which had an even distribution of particles on them so that random microscope fields could be taken which would be statistically significant. Although many data points fell on the low end of the curve, there are few data points for the high turbidity water (greater than about $E/E_0 = 8$).

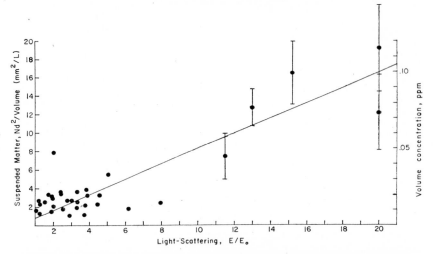

Figure 16   Surface area concentration of particles recovered from water samples (Nd^2/L) versus light-scattering (E/E_0) measured at the depth of the sample. For a mean spherical particle diameter of 12 μ, the concentration is given in ppm on the right-hand side. The estimated error in the count analysis is indicated for the samples on the critical high end

Although there is considerable scatter of data, there is clearly a direct relationship between the two plotted variables. The scatter is not surprising, considering the errors involved in the particulate matter concentration determinations and the possible optical errors in the concentration-scattering relation.

For purposes of deriving particulate matter concentrations from the light scattering intensities, the line in Figure 16 with a slope of 0.84 will be used. Using a direct proportionality line allows integrated light scattering values to be converted directly to volumes of particulate matter. Volume concentrations in ppm are derived from the Nd^2 values using an assumed spherical particle size of 12 μ, the mean of the mean sizes (by volume) of all samples in the nepheloid layer (Figure 15). The assumption of sphericity will result in a ppm concentration too large if the particles are predominately flaky in shape or too small if the particles are predominantly cubic or tabular.

Concentrations in the nepheloid layer are on the order of 0.01 to 0.1 ppm. This compares with the average figure from Jacobs and Ewing (1969) of 0.07 mg/L or 0.035 ppm (using a density of 2 g/cm^3 to convert from weight to volume concentration) for the deep nepheloid waters of the Northwest Atlantic. The light scattering measurements of Jerlov indicate concentrations of 0.050 mg/L or 0.025 ppm in the deep western equatorial Atlantic in water which he concludes is Antarctic Bottom Water (Jerlov, 1968; Figure 69).

Folger (1968) in a study of suspended matter in surface water found total concentrations averaging 0.05 mg/L (0.025 ppm) in the Atlantic equatorial regions and 0.1 mg/L (0.05 ppm) along a section at about 40° N. Of this total, most was living plankton, and less than 10% was material such as silts and clays, carbonate and siliceous organisms which might eventually contribute to the deep nepheloid waters. The concentration of carbonate-free mineral matter was about 0.3% of the total. Manheim *et al.* (1970) found surface water suspended matter concentrations in off-shore regions less than 0.1 mg/L (0.05 ppm), less than 10% of which was non-combustible mineral matter. Folger (1968) considered the relationship of these extremely low concentrations of surface water mineral particles to the observed sedimentation rates of clays on the seafloor and concluded that the latter could not be accounted for by fallout from surface waters. Horizontal transport of the clays and silts by nepheloid bottom currents is the likely mechanism of their distribution. For comparison, typical concentrations of atmospheric particulate matter in mid-ocean regions are on the order of 10^{-8} ppm (Parkin *et al.*, 1970). Unless the particle residence time differs by a factor of 10^7 between atmosphere and ocean, the atmospheric contribution to the nepheloid layer must be very small.

The increase in density of water due to the suspended matter in the nepheloid layer is insignificant. For a concentration of 0.1 ppm the density increase is 0.1×10^{-6} for a particle density of 2 g/cm^3. This concentration of particles would alter the density as much as a temperature decrease of 0.0006 °C. This is too small to have any significant effect on the density relationship of nepheloid to non-nepheloid waters. This same conclusion was reached by Hunkins *et al.* (1969) for a bottom nepheloid layer observed in the Arctic basin.

VOLUME OF SUSPENDED SEDIMENT IN THE BASIN

Using Figure 11 total quantities of suspended matter can be derived for the whole basin or for parts of it. The areas under the E/E_0 versus depth curves, $\int E/E_0 \, dD$ are units of meters, since the scattering units are a dimensionless ratio. If we assume proportionality between E/E_0 and concentration (Figure 16), a conversion can be made from values of $\int E/E_0 \, dD$ to volumes of suspended matter per column of ocean water of unit cross section. Such an average value for a given region of the basin, multiplied by the area of that region, will give the total volume of suspended matter in that region.

The line in Figure 16 represents the relation:

$$\text{Concentration (ppm)} = 0.005 E/E_0,$$

on the assumption of a mean spherical particle size of $12 \, \mu$ (Figure 15). Multiplying the $\int E/E_0 \, dD$ values by 0.005, gives an estimate of the volume of particulate matter in cubic centimeters per square meter vertical column of ocean water through the nepheloid layer. A one degree grid was placed over the map and a value of $\int E/E_0 \, dD$ was assigned to each degree square using the contours of Figure 11. These values were summed up for all square degrees in the basin in groups of 5 degrees latitude and each sum multiplied by 0.005 to convert to ppm. meters or cm^3/m^2. This value was then multiplied by the number of meters^2 per square degree, in each latitudinal group. The eastern edge of the basin was defined as the boundary between upper and middle step of the Mid Atlantic Ridge (Heezen and Menard, 1961). The estimated total volume of particulate matter in suspension derived in this manner is 1.2×10^{14} cm^3 or about 10^8 metric tons.

If the active fallout rates of this suspended sediment were known, then a residence time of sediment in suspension could be derived by dividing the volume in suspension by this fallout rate. However only net sedimentation rates over thousands of years are known and if there is active erosion from the sea floor then this net sedimentation rate is not equal to the fallout rate but to the difference between fallout and erosion rates. At present there is no way of knowing whether the sediment goes through more than one suspension-deposition cycle of if so, how many cycles. Eittreim (1970) derives an average uncompacted lutite sedimentation rate of 1.9 cm/10^3 yrs from the published reports on piston cores and profiler data in the basin and further relates this to the volume in suspension. If it is assumed that no resuspension occurs, a maximum residence time in suspension of about 1 yr is derived. Since the fallout rate will be increased above the observed sedimentation rate by a factor equal to the number of resuspension cycles which the particles are subject to, the residence time will be decreased by this same factor.

A schematic model of the North American Basin nepheloid layer is shown in Figure 17 with the inputs to and outputs from the nepheloid layer. The inputs and outputs might be combined into a theoretical budget, but the

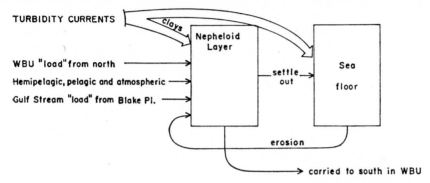

Figure 17 Hypothetical scheme for sediment inputs to and outputs from the nepheloid layer, showing five main sources of sediment on the left is decreasing importance from top to bottom. See text for explanation

number of undetermined variables prevents any unique solution to such a budget. What are believed to be the four major sources are indicated in the figure for the following reasons:

1   The great intensification of the nepheloid layer in the region of the continental rise between Cape Cod and Cape Hatteras (the most intense light scattering values, E/E_0 as high as 20, are found here) suggests that large amounts of material are being added to the water column here. The fact that in this area, the adjacent continental slope is intensely cut with canyons further suggests that this material is being added to the water column by turbidity currents, either of the catastrophic, high velocity type produced by sudden slumping of unstable sediments, or a slower more or less steady-state type of the sort discussed by Moore (1966) and Shephard et *al.* (1969) off the west coast of the U.S. Such turbidity currents might be active all along the continental slope from the Grand Banks to Cape Hatteras, but the intensification of the light scattering in the region off New York suggests that such activity is most prominent in the latter region.

2   The *Vema* 23 stations in the northern part of the basin (Jones *et al.*, 1970) show that the WBU, upon entering the basin, is laden with a considerable amount of suspended sediment, accounting for one major source.

3   A continuous rain of surface water organisms, along with atmospheric dust and possibly surface water-transported terrigenous detritus seems

likely on the basis of the presence of planktonic organic remains in the nepheloid water samples. For the reasons outlined previously, especially the relatively low concentrations of clays and silts found by Folger (1968) in mid-ccean surface waters, this is not thought to be a primary source.

4 It is likely that the Gulf Stream is transporting suspended sediment from continental sources in southeastern U.S. (Pratt and McFarland, 1966) while at the same time scouring sediments from the Blake Plateau surface where many pre-Pleistocene sediments have been exposed (Ewing *et al.*, 1965). These suspended sediments might then settle to the deep and bottom waters after passing over the edge of the Blake Plateau. The downward settling of these sediments into the WBU may account for, possibly along with the upward diffusion of the deeper suspended sediments, the increased thickness of the nepheloid layer south of the northern Blake Plateau.

The data available are insufficient to estimate the relative amounts of all these inputs, but it may be instructive to compare the output from the nepheloid layer via sedimentation on the basin floor with estimated inputs and outputs via the WBU.

The southward transport of the WBU at 30° N along the continental margin was estimated as 10 million m^3/sec by Barrett (1965) and 20 million m^3/sec by Amos *et al.* (1970). The latter figure probably more closely represents the real WBU transport since it was obtained from a section which probably enccmpassed the whole zone of WBU flow whereas Barrett's extended only to 3500 m depth and may have missed much of the transport. No data are available on transport along the northern edge of the basin, south of the Grand Banks or along the southwest margin of the basin off the Bahama Islands. However, where it exits from the Labrador Sea at 51° N, Swallow and Worthington (1969) calculate the transport of this bottom current to be about 10 million m^3/sec. Just beyond the southern end of the basin, near latitude 15° N, bottom velocities in the WBU have been estimated to be 2–3 cm/sec (McCoy, 1969) as compared to bottom velocities of 10 to 20 cm/sec in the basin to the north. The net southern output therefore must be fairly small and would be further diminished by considering the amount of sediment brought into the basin from the south by the AABC. As a rough estimate, assume that 10 million m^3/sec is transported into the northern end of the basin, and that 5 million m^3/sec is transported out at the southern end. From Figures 11 and 16, average concentrations are derived for a nepheloid layer of 1 km thickness at the WBU input in the north and a nepheloid layer thickness of 2 km at the WBU outflow in the south. This gives 0.9×10^{13} cm^3/yr and 0.3×10^{13} cm^3/yr of sediment transported into and out of the basin respectively. Compare this with 12.3×10^{13} cm^3/yr deposited on the basin floor (Eittreim, 1970).

Estimates of suspended sediment transport in the bottom water have been made in the Argentine Basin and it was concluded that the major source of sediment was the adjacent continent (Ewing *et al.*, 1971). The analogy of the North American Basin to the Argentine Basin is striking, with both basins having the following similarities:

1   Vigorous bottom water circulation.
2   A nepheloid layer comprising the bottom one kilometer of the water column.
3   Large bodies of lutite sediments, in places being formed into dune structures or depositional ridges, such as the Blake-Bahama Outer Ridge in the North American Basin and the Zapiola Ridge in the Argentine Basin.
4   Adjacent continental slopes intensely dissected by canyons.

The similarities in sedimentation processes in these basins and their uniqueness on a global scale are probably due to the fact that these are the only major basins of the world, aside from possibly the circumantarctic, in which there are both vigorous bottom currents and an ample supply of sediments.

As noted in the hypothetical model, Figure 17, the emphasis is on turbidity currents as a major source of the nepheloid layer particles for the reasons already outlined. Resuspension of sediments from the bottom in significant quantities is also a tenable source. If such recycling of sediments occurs, a method must be found to determine the rates of recycling in order to develop a more realistic model.

INTENSITY OF TURBULENCE IN THE DEEP WATER

Many studies have been made of turbulence in streams and rivers and its relationship to suspended matter carried (Schmidt, 1917; Leighly, 1933; O'Brien, 1933; Hjulstrom, 1935). In a turbid water column with an equilibrium turbidity profile a balance exists between upward diffusion of the particles and their downward settling by gravity. The first basic assumption that must be made in order to infer intensity of turbulence from data on vertical gradients of suspended matter is that of an equilibrium turbidity profile. This assumption implies that the particle concentration at a given depth does not change appreciably (a) with time and (b) with distance in a horizontal "downstream" direction. Over the years that measurements have been taken in this basin (1964 to the present) the nepheloid layer has consistently been observed. The data presented here span the time from January 1966 to April, 1968. In the group of stations off New York, *Conrad* 11 stations were taken in November, 1967, the *Vema* 25 stations in April,

1968, and the *Gibbs* 68-1 stations in November, 1968. The similarity of these data (Figure 9) suggests a steady-state phenomena. The *Conrad* 8 data presented by Ewing and Thorndike (1965) were taken in 1964 in the same area and show a layer of similar thickness and although the light scattering values cannot be reduced to the same units as those used in this study, the intensity appears to be approximately the same.

In the presence of a diffusing mechanism such as turbulence there will be a net transport of any property which exhibits concentration gradients in space. At a given instant the turbulent water which is moving in the direction of negative gradient contains a higher concentration of the property than the water moving in the opposite direction. Hence, the net turbulent transport or so-called eddy diffusion of the property is towards regions of lower concentration. The equation describing this eddy diffusion flux in the vertical direction is:

$$F_z = -A_z \frac{dc}{dz},$$

where F_z = vertical turbulent transport of the property in question.

A_z = coefficient of vertical mixing (Austausch coefficient, Schmidt 1917, or coefficient of eddy diffusion).

c = concentration of property (e.g. salinity, suspended matter, etc.).

z = vertical distance from bottom.

The negative sign implies that the direction of net diffusion is opposite to the direction of positive concentration gradient.

Suspended particles with negative buoyancy will settle downward and their flux will be wc, where w is particle settling velocity. Thus the net vertical flux is:

$$F_z = -A_z \frac{dc}{dz} - wc$$

Assuming a balance between upward diffusion flux and downward gravitational flux, F_z is set equal to zero and

$$\frac{w}{A_z} = -\frac{1}{c}\frac{dc}{dz} = -\frac{d(\ln c)}{dz}$$

It can be seen from the last equation that if settling velocities are in cm/sec and vertical distance in cm, A_z will be in units of cm²/sec. Also note that the results are not dependent on the concentration units used, since these cancel. Thus from the slope of the log-concentration versus depth curve, ratios of

A_z/w may be obtained directly. Ichiye (1966) obtained such ratios for two nephelometer stations taken on *Conrad* 8 on the lower continental rise off New York and obtained values of A_z/w ranging from 0.2×10^4 to 12×10^4 cm. For reasonable particle settling velocities (Sverdrup *et al.*, 1942, Table 105), these give maximum A_z values in the range of 10 to 10^3 cm^2/sec. By assuming A_z approached zero at the bottom, Ichiye forced the conclusion of a maximum in A_z at 100 m. This 100 m level was rather arbitrary, being based on the theoretical constraint of zero vertical motion at the bottom. Although such a boundary layer must exist, it appears that its dimensions are on order of meters or less, too small to be resolved with the nephelometer.

The assumption that at any given station the nephelometer profile represents an equilibrium profile of turbidity may be a dangerous one, since at a given location in the basin, at a given time, time-dependent and lateral distance dependent perturbations of the profile may exist. The fact that on some stations a change in sign of dc/dz occurs over small depth intervals (cf. Figure 3) shows that in fact simple equilibrium between upward diffusion flux and downward gravitational flux does not hold everywhere. However, if a gradient is constructed which is an average of many stations over a wide region, such an assumption for that gradient is probably a safe one. For the purpose of constructing such averages in Figure 18 the nephelometer curves are grouped geographically and it can be seen that each group has a different characteristic shape. Possible explanations for the occurrence of these different types over broad areas are discussed in Eittreim *et al.* (1969).

Stokes' law is used to derive settling velocities for three different size particles. It is generally agreed that Stokes' formula holds for particles smaller than about 50 μ (Gripenberg, 1939). Although the formula is derived for spherical particles, deviations from sphericity do not cause large errors (Krumbein and Pettijohn, 1938) The choice of particle density is difficult. The maximum density which might be encountered in the ocean is about 2.7 (quartz density = 2.65). The clay minerals which predominate in the nepheloid layer are hygroscopic, decreasing their densities to an average of about 2.0 (Grim, 1953, table 42).

In the analysis of particulate matter recovered from water samples, aggregates have been considered as particles in the counts. This is proper, for such aggregates have properties of light scattering and settling velocity very different from those of the individual particles of which they are made up. These aggregates constitute from zero to 60% of the total particles and average about 25%. Their density would be affected by the approximately 50% water-filled pore space between particles. The density of these aggregates is difficult to estimate but is probably close to 1.5. Organic particles would have densities on the average probably only slightly higher than water. As a best estimate, considering the water-carrying clay minerals, the aggregates and the biological particles, 1.8 will be assumed as an average particle density, keeping in mind that this figure may be uncertain by as much as 40%.

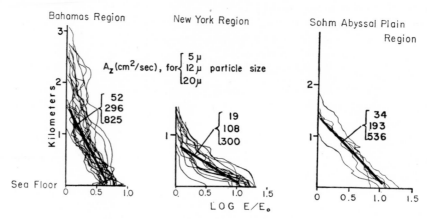

Figure 18 Nepheloid layer portions of the nephelometer curves in the North American Basin placed in three groups and derived eddy diffusivities. The bold lines are average gradients constructed by eye. The eddy diffusivities shown are calculated from the average gradients for three assumed particle sizes with assumption of particle density of 1.8 g/cm³

It is assumed that light scattering values, E/E_0, are proportional to concentration of particulate matter (Figure 16). An average gradient of the bottom water light scattering curves for each of the three groups was chosen by eye. Using the average gradient for each group of curves, A_z values are calculated for 5, 12 and 20 μ diameter particles, respectively and the results shown in Figure 18. On the basis of the particle size distribution found in this study and illustrated in Figure 15, a diameter of 12 μ is the preferred value.

The values of A_z derived are at least an order of magnitude higher than previously found for vertical mixing coefficients in deep waters. Sverdrup *et al.*, 1942, Table 65, give A_z values of various workers computed from time and space variations of salinity, temperature and oxygen measurements ranging from 0.02 to 320 cm²/sec with no values larger than 10 for deep waters. Broecker *et al.* (1967) estimate a value of 50 cm²/sec in the Argentine Basin, on the basis of the distribution of excess radon near the ocean bottom.

Methods of calculation of A_z usually depend on strong gradients of some conservative property between the water masses involved. Obviously where the mixing is very thorough, no such strong gradients exist and the method is difficult or impossible to employ. Suspended matter on the other hand, by virtue of its gravitational settling, maintains a measurably strong gradient despite the turbulence, and therefore can be used for measurement of this mixing coefficient. Obviously more precise measurements of particle concentration, size and density are needed to determine A_z values with precision but the conclusion that eddy diffusivities are abnormally high in these waters is inescapable.

If an overestimate of 40% was made in the assignment of an average particle density, this would decrease the A_z determination by a factor of two. Assuming such an overestimate, the values of A_z are still high, relative to other published values. The cause of this turbulence must be shear due to the presence of bottom currents, the WBU and AABC, active over the deep areas of the basin, especially along the western side. These bottom currents, which represent a large amount of kinetic energy, must, in conjunction with the near-neutral stability of the water, result in the upward diffusion of particles observed.

Ewing and Connary (1970), comparing the light scattering data in the North Pacific with profiles of temperature, salinity and oxygen, concluded that the bottom waters there are extremely well-mixed to an average height of about one kilometer above the bottom. Also, Hunkins *et al.* (1970) concluded that the bottom nepheloid layers in parts of the Arctic Ocean are the result of turbulence in the bottom few hundred meters of the water column.

CONCLUSIONS

1   The North American Basin nepheloid layer is a permanent feature of the bottom waters covering the continental rise, abyssal plains and Bermuda Rise. On the average, a ten-fold increase in light scattering occurs from the clear water at mid-depths to the maximum intensity scattering near the bottom in the nepheloid layer.

2   On the basis of (a) the general conformity of the "top" of the nepheloid layer to the deep water isotherms and (b) the occurrence of the nepheloid layer in regions where bottom currents are strongest, it is concluded that the nepheloid layer is related to the major bottom currents of the North American Basin, the Western Boundary Undercurrent and the Antarctic Bottom Current, the former being identified with the highest scattering intensities.

3   Particulate matter recovered in water samples in the nepheloid layer shows the layer to be composed of clay and silt sized mineral matter, organic debris and silicate and carbonate tests, in generally that order of abundance. The highest concentrations recovered were in regions of the most intense light scattering and were slightly higher than 0.1 ppm (0.2 mg/L for a particle density of 2.0). Average particle size, including aggregates, for the nepheloid layer is 12 μ by volume.

4   Turbulence associated with the WBU and AABC combined with an influx of sediments from the adjacent continent is responsible for the nepheloid layer. The relative homogeneity of the bottom waters in the North American Basin, i.e. the lack of strong density gradients, reflects this turbulence. Coefficients of vertical eddy diffusion average about 10^2 cm^2/sec, on the basis of an assumed balance between downward gravitational settling of particles and their upward turbulent diffusion.

5   The highest intensity light scattering occurs near the bottom in the region of the continental rise between Cape Hatteras and Cape Cod where there are numerous submarine canyons. This suggests that the major source of suspended matter is terrigenous detritus channeled in these canyons on the continental slope and rise. The mode of transport of this material in the canyons is most likely gravity-controlled turbidity currents. Tidal action and transient surges on the shelf might also be responsible for delivery of suspended sediments to the canyons.

Acknowledgments

E.M.Thorndike provided helpful advice and criticism on data reduction procedures and instrumental problems. We benefitted greatly from the advice of D.Folger in the planning and carrying out of the water sampling program and from W.B.F.Ryan in the computer programming stages. Allan Bé and Oswald Roels kindly allowed use the use of laboratory equipment for the particulate matter analysis. A.Gordon and M.Langseth critically reviewed the manuscript. This research was supported by the National Science Foundation under grants GA 1615 and GA 550 and the Office of Naval Research contract no. N-00014-67-A-0108-0004.

References

Amos, A., A.Gordon, and E.D.Schneider, Water masses and circulation patterns in the region of the Blake Bahama outer ridge, *Deep-Sea Res.*, **18**, 145–165, 1971.

Barrett, J.R., Subsurface currents off Cape Hatteras, *Deep-Sea Res.*, **12**, 173–184, 1965.

Beardsley, G.F., H.Pak, K.Carder, and B.Lundgren, Light scattering and suspended particles in the eastern equatorial Pacific Ocean, *J. Geophys. Res.*, **75**, 2837–2845, 1970.

Broecker, W.S., Y.H.Li, and J.Cromwell, Radium-226 and Radon-222: Concentration in Atlantic and Pacific oceans, *Science*, **158**, 1307–1310, 1967.

Connary, S.D., Nepheloid layer and bottom circulation in the eastern basins of the South Atlantic, Masters Thesis, Columbia Univ., New York, 24 pp., 1970.

Dietrich, G., Some thoughts on the working-up of the observations made during the "Polar Front Survey" in the IGY 1968, *Rapp. Proc.-Verb. Reun. Cons. Perm. Int. Explor. Mer*, **149**, 103–110, 1961.

Einstein, H.A. and R.B.Krone, Experiments to determine modes of cohesive sediment transport in salt water, *J. Geophys. Res.*, **67**, 1451–1461, 1962.

Eittreim, S., Suspended matter in the deep waters of the northwest Atlantic Ocean, Ph. D. Thesis, Columbia Univ., New York, 166 pp., 1970 (available on microfilm).

Eittreim, S., M.Ewing, and E.M.Thorndike, Suspended matter along the continental margin of the North American basin, *Deep-Sea Res.*, **16**, 613–624, 1969.

Eittreim, S., A.L.Gordon, E.M.Thorndike, and P.Bruchhausen, The nepheloid layer and observed bottom currents in the Indian-Pacific Antarctic sea, this volume.

Ericson, D.B., M.Ewing, G.Wollin, and B.C.Heezen, Atlantic deep-sea sediment cores, *Bull. Geol. Soc. Am.*, **72**, 193–286, 1961.

Ewing, J., M.Ewing, and R.Leyden, Seismic-profiler survey of Blake plateau, *Bull. Am. Assoc. Petrol. Geologists*, **50**, 1948–1971, 1966.

Ewing, M. and S.Connary, Nepheloid layer in the North Pacific, *Geol. Soc. Am. Memoir*, No. 126, J.D.Hays, ed., 41–82, 1970.

Ewing, M., S.Eittreim, J.Ewing, and X. Le Pichon, Sediment transport and distribution in the Argentine basin: Part 3, Nepheloid layer and processes of sedimentation, in *Physics Chemistry Earth*, **8**, 49–77, 1971.

Ewing, M. and J.Ewing, Distribution of oceanic sediments, in *Studies on Oceanography*, K.Yoshida, ed., Tokyo Univ. Press, Tokyo, 525–537, 1964.

Ewing, M. and E.M.Thorndike, Suspended matter in deep ocean water, *Science*, **147**, 1291–1294, 1965.

Folger, D.W., Trans-Atlantic sediment transport by wind, Ph. D. Thesis, Columbia Univ., New York, 189 pp., 1968.

Fox, P.J., W.C.Pitman III, and F.Shephard, Crustal plates in the central Atlantic: Evidence for at least two poles of rotation, *Science*, **165**, 487–489, 1969.

Fuglister, F.C., Atlantic Ocean atlas of temperature and salinity profiles and data from the International Geophysical Year 1957-58, *Atlas Ser. Woods Hole Oceanogr. Inst.*, **1**, 1960.

Gordon, D.C., Some studies on the distribution and composition of particulate organic carbon in the North Atlantic Ocean, *Deep-Sea Res.*, **17**, 233–243, 1970.

Grim, R.E., *Clay Mineralogy*, McGraw-Hill, New York, 348 pp., 1953.

Gripenberg, S., Mechanical analysis, in *Recent Marine Sediments*, P. D. Trask, ed., Am. Assoc. Petrol. Geologists, 532–557, 1939.

Heezen, B.C. and C.Hollister, Deep-sea current evidence from abyssal sediments, *Marine Geol.*, **1**, 141–174, 1964.

Heezen, B.C., C.D.Hollister, and W.F.Ruddiman, Shaping of the continental rise by deep geostrophic contour currents, *Science*, **152**, 502–508, 1966a.

Heezen, B.C. and H.W.Menard, Topography of the deep-sea floor, in *The Sea*, 3, M.N. Hill, ed., Interscience, New York, 233–280, 1963.

Heezen, B.C., E.D.Schneider, and O.H.Pilkey, Sediment transport by the Antarctic Bottom Current on the Bermuda Rise, *Nature*, **211**, 611–612, 1966b.

Hesselberg, T., Über die Stabilitätsverhältnisse bei vertikalen Verschiebungen in der Atmosphäre und im Meer, *Ann. Hydrogr. Berl.*, 118–129, 1918.

Hjulstrom, F., Studies of the morphological activities of rivers as illustrated by the River Fyris, *Bull. Geol. Inst. Univ. Upsala*, **25**, 221–527, 1935.

Hollister, C.D., Sediment distribution and deep circulation in the northwest Atlantic, Ph. D. Thesis, Columbia Univ., New York, 472 pp., 1967.

Horne, R.A. and D.S.Johnson, The viscosity of water under pressure, *J. Phys. Chem.*, **70**, 2182–2190, 1966.

Hunkins, K., E.M.Thorndike, and G.Mathieu, Nepheloid layers and bottom currents in the Arctic Ocean, *J. Geophys. Res.*, **74**, 6995–7008, 1969.

Ichiye, T., Turbulent diffusion of suspended particles near the ocean bottom, *Deep-Sea Res.*, **13**, 679–685, 1966.

Jacobs, M.B. and M.Ewing, Suspended particulate matter: Concentration in the major oceans, *Science*, **163**, 380–383, 1969.

Jerlov, N.G., The particulate matter in the sea as determined by means of the Tyndall meter, *Tellus*, **7**, 218–225, 1955.

Jerlov, N.G., *Optical Oceanography*, Elsevier, Amsterdam, 194 pp., 1968.

Jones, E.J.W., M.Ewing, J.Ewing, and S.L.Eittreim, Influences of Norwegian Sea Overflow Water on sedimentation in the northern North Atlantic and Labrador Sea, *J. Geophys. Res.*, **75**, 1655–1680, 1970.

Knauss, J., A technique for measuring deep ocean currents close to the bottom with an unattached current meter and some preliminary results, *J. Mar. Res.*, **23**, 237–245, 1965.

Krumbein, W.C. and F.J.Pettijohn, *Manual of Sedimentary Petrography*, Appleton-Century-Crofts, Inc., New York, 549 pp., 1938.

Kuenen, Ph.H., Experiments in connection with turbidity currents and clay-suspension, *Proc. Symp. Colston Res. Soc.*, **17**, 47–74, 1965.

Leighly, J., Turbulence and the transportation of rock debris by streams, *Geogr. Rev.*, **24**, 453–464, 1934.

Lynn, R.J. and J.L.Reid, Characteristics and circulation of deep and abyssal waters, *Deep-Sea Res.*, **15**, 577–598, 1968.

Manheim, F.T., R.H.Meade, and G.C.Bond, Suspended matter in surface waters of the Atlantic continental margin from Cape Cod to the Florida Keys, *Science*, **167**, 371–376, 1970.

Matthews, D.J., *Tables of the Velocity of Sound in Pure Water and Sea Water*, Hydrographic Dept., Admiralty, London, 52 pp., 1939.

McCoy, F.W., Bottom currents in the western Atlantic Ocean between Lesser Antilles and the mid-Atlantic ridge, *Deep-Sea Res.*, **16**, 179–184, 1969.

Moore, D.G., Structure, litho-orogenic units and post-orogenic basin fill by reflection profiling: California continental borderland, Ph. D. thesis, Univ. Groningen, Netherlands, 151 pp., 1966.

Niskin, S.J., A water sampler for microbiological studies, *Deep-Sea Res.*, **9**, 501–503, 1962.

O'Brien, M.P., Review of the theory of turbulent flow and its relation to sediment transportation, *Trans. Am. Geophys. Union*, **14**, 487–491, 1933.

Parkin, D.W., D.R.Phillips, R.A.L.Sullivan, and L.Johnson, Airborne dust collections over the North Atlantic, *J. Geophys. Res.*, **75**, 1782–1793, 1970.

Peterson, M.N.A., N.T.Edgar, M.Cita, S.Gartner, R.Goll, C.Nigrini, and C. von der Borch, *Initial Reports of the Deep Sea Drilling Project*, Volume 2, U.S. Government Printing Office, Washington, D.C., 501 pp., 1970.

Pratt, R.M. and P.F.McFarland, Manganese pavements on the Blake plateau, *Science*, **151**, 1080–1082, 1966.

Rowe, G.T. and R.J.Menzies, Deep bottom currents off the coast of North Carolina, *Deep-Sea Res.*, **15**, 711–719, 1968.

Ryan, W.B.F., The floor of the Mediterranean Sea, Ph. D. thesis, Columbia Univ., New York, 404 pp., 1971.

Schmidt, W., Wirkungen der ungeordneten Bewegung im Wasser der Meere und Seen, *Ann. Hydrogr. Marit. Meteorol.*, **45**, 367–445, 1917.

Schneider, E.D., P.J.Fox, C.D.Hollister, H.D.Needham, and B.C.Heezen, Further evidence of contour currents in the western North Atlantic, *Earth Planet. Sci. Lett.*, **2**, 351–359, 1967.

Shepard, F.P., R.F.Dill, and U. von Rad, Physiography and sedimentary processes of La Jolla submarine fan and fan-valley, California, *Bull. Am. Assoc. Petrol. Geologists*, **53**, 390–420, 1969.

Stefansson, U., Dissolved nutrients, oxygen and water masses in the northern Irminger Sea, *Deep-Sea Res.*, **15**, 541–575, 1968.

Stommel, H., A survey of ocean current theory, *Deep-Sea Res.*, **4**, 149–184, 1957.

Sverdrup, H.U., M.W.Johnson, and R.H.Fleming, *The Oceans*, Prentice-Hall, Inc., Englewood Cliffs, N.J., 1087 pp., 1942.

Swallow, J.C. and L.V.Worthington, An observation of a deep countercurrent in the western North Atlantic, *Deep-Sea Res.*, **8**, 1–9, 1961.

Swallow, J.C. and L.V.Worthington, Deep currents in the Labrador Sea, *Deep-Sea Res.*, **16**, 77–84, 1969.

Thorndike, E.M. and M.Ewing, Photographic nephelometers for the deep sea, in *Deep-Sea. Photography*, J.B.Hersey, ed., Johns Hopkins Press, Baltimore, 113–116, 1967.

Volkmann, G., Deep current observations in the western North Atlantic, *Deep-Sea Res.*, **9**, 493–500, 1962.

Whitehouse, U.G., L.M.Jeffrey, and J.Debbrecht, Differential settling tendencies of clay minerals in saline waters, *Clays and Clay Minerals, Proc. Seventh Nat. Conf.*, **5**, 1–79, 1960.

Wüst, G., Das Bodenwasser und die Gliederung der Atlantischen Tiefsee, *Wiss. Ergeb. dt. atlant. Exped. Meteor 1925–1927*, **6** (1), 1–107, 1933.

Wüst, G., Schichtung und Zirkulation des Atlantischen Oceans, *Wiss. Ergeb. dt. atlant. Exped. Meteor 1925–1927*, **6**, (1) (2), 108–288, 1935.

Wüst, G., Stromgeschwindigkeiten im tiefen und Bodenwasser des Atlantischen Ozeans auf Grund dynamischer Berechnung der Meteorprofile der Deutschen Atlantischen Expedition 1925–27, Papers in Marine Biology and Oceanography, Suppl. *Deep-Sea Res.*, **3**, 373–397, 1955.

The Nepheloid Layer and Bottom Circulation in the Guinea and Angola Basins*

STEPHEN DODD CONNARY
AND MAURICE EWING

Lamont-Doherty Geological Observatory of Columbia University
Palisades, New York 10964

Abstract A nepheloid layer less than 200 m thick is present in the eastern Guinea Basin; in the Angola Basin, the nepheloid layer is 500 to 1500 m thick and may be correlated with the homogeneous bottom water. A thicker layer is found in areas of higher topographic relief. In the deep part of both basins, the scattering intensity of the most turbid water of the nepheloid layer is about 1.5 times greater than that in the clearest water in the column. Compared to the Guinea and Angola Basins, scattering intensities are greater in the western North Atlantic by a factor of about 6, in the western South Atlantic and Cape Basin by about 4, and are roughly comparable to those in the North Pacific. One or more prominent light scattering zones above the main nepheloid layer is typical at stations on the continental slope of the Guinea and Angola Basins, but these zones are weak or absent in the deep basins.

Recent temperature and salinity data broadly support the earlier conclusions on deep circulation in these basins. Anomalous cooling of the bottom water in the southwestern Angola Basin suggests that there may be deep gaps in the Walvis Ridge through which the colder Cape Basin bottom water flows. Nephelometer data also support this conclusion. The northward spreading of this influxing Cape Basin water may be blocked by a wide area of high-relief topography to the north of the Walvis Ridge.

I INTRODUCTION

Wüst (1933) studied in detail the horizontal spreading of Antarctic Bottom Water (AABW) in the Atlantic Ocean. He observed that in the western trough of the South Atlantic there exists a north-flowing bottom water mass of Antarctic origin whose upper boundary is marked by a strong tempera-

---

* Lamont-Doherty Geological Observatory Contribution Number 1617.

ture and salinity gradient and whose properties are observable to at least 40° N in the western North Atlantic. In contrast, the eastern trough of the South Atlantic, which is divided into the Guinea, Angola, and Cape Basins, exhibits no such pattern of bottom circulation. It was Wüst's belief that the AABW in the Cape Basin is completely blocked from further northward advance into the eastern South Atlantic by the high Walvis Ridge which separates the Cape and Angola Basins near 25° S. The bottom waters of the Guinea and Angola Basins north of the Walvis Ridge flow from the Brazil Basin through the Romanche Fracture Zone, a deep gap in the Mid-Atlantic Ridge near the Equator.

Wüst (1957) later calculated the geostrophic currents from the *Meteor* profiles in the South Atlantic. The resulting distribution of north-south current components corresponds closely with his prior conclusions based on his "core" method.

In 1963 a program of measuring *in situ* light scattering in the deep sea was initiated at Lamont-Doherty Geological Observatory using a photographic nephelometer designed by Thorndike and Ewing (1967). A major result of this world-wide study was the detection of a nepheloid layer extending several hundred meters above the ocean floor in which there is a relative increase in the scattering of light with depth (Ewing and Thorndike, 1965; Ewing et al., in press; Eittreim et al., 1969; Hunkins et al., 1970; Ewing and Connary, 1971; Eittreim and Ewing, this volume; Eittreim et al., this volume). This scattering has been attributed to small particles of mineral and organic material which, it is believed, are kept in suspension by turbulence resulting from the interaction of abyssal currents and the sea floor topography. Attempts have been made to relate particle mass concentration to the intensity of light scattering (Eittreim, 1970; Eittreim and Ewing, this volume). Good correlations have been lacking largely because light scattering is a function of particle size, refractive index, absorption, and shape, as well as particle concentration. Other variables being constant, a large number of small particles are expected to scatter more light than a few larger ones with equal mass concentration. These relations have been observed by Jacobs et al. (in prep.) who counted four to five times more particles in the very turbid water of the North American Basin than in the relatively clear intermediate water over the Mid-Atlantic Ridge. They also concluded that the most abundant particles in both the clear and nepheloid water are non-opaque mineral grains less than two microns in diameter. A slightly higher percentage of these particles was found in the nepheloid water.

For the present study nephelometer records from the Cape, Angola, and Guinea Basins were examined (Figure 1). The features common to all the records are the following: high scattering in the surface layer, commonly about 300 m thick; in the intermediate water scattering decreases slowly with depth, to a level at which a pronounced scattering increase begins and continues to the sea floor. The depth at which this significant increase commences is taken as the top of the nepheloid layer. Light scattering intensity in the nepheloid layer is expressed as a ratio E of film exposure near the sea floor to exposure in the clearest water at intermediate depths. Within the nepheloid zone, film exposure is always greatest near the sea floor, while in the basins under study, scattering at the sea floor is generally less than that in the surface layer.

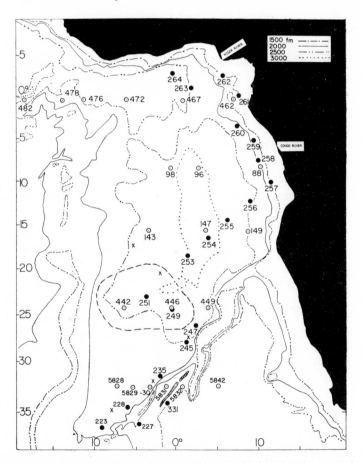

Figure 1 Location chart of nephelometer stations (solid circles) and *Crawford* and *Atlantis* hydrographic stations (open circles) in the Guinea, Angola, and Cape Basins. Nephelometer stations at which relative film exposure could not be calculated are indicated by X's. The region enclosed by the dashed line in the southern Angola Basin outlines an area of rugged topography discussed in the text. Soundings are in fathmos; 500 fathmos equal 915 m

As the present nepholometer technique does not measure absolute light scattering, one must choose a suitable reference to which measured intensities may be related. It is assumed that, over a limited geographic area, the clearest water has approximately the same intensity of light scattering. This hypothesis permits a comparison of scattering intensity at two or more stations within the area. As most of the suspended particulate matter is concentrated at the bottom of the nepheloid zone, it seems most reasonable to expect the greatest variations to occur here, while the clearest water remains to a first approximation laterally uniform.

II ANGOLA BASIN

Fifteen light scattering profiles (Figure 2) are in the Angola Basin as shown in Figure 1. The locations of several other profiles in the basin are indicated for which the calculation of relative film exposure to the desired accuracy is not possible. These profiles are not shown, but in every respect they have the same features as nearby stations. For the profiles in Figure 2, only the main nepheloid layer and the scattering zones at intermediate depths where present are shown.

The ten nephelometer stations in the deep part of the Angola Basin and two from the adjacent part of the Cape Basin may be described in three groups according to the depth of the top of the nepheloid layer. These are the following:

Group	Stations	Top of Layer	Average E
1	331, 227, 223, 228, 235	3.5 km	2.1
2	245, 247, 249, 251	4.1 km	1.4
3	253, 254, 255	4.9 km	1.3

One may also group these stations according to the scattering intensity ratio E. The main nepheloid layer at Group I stations has nearly similar intensitity; intensities in Groups 2 and 3 are all very similar but are considered significantly weaker than those in Group 1.

The five remaining Angola Basins stations are all on the eastern boundary of the basin in water depths of 2.5 to 4.5 km. Each of these shows a nepheloid layer at the bottom, generally stronger (average E about 2.2) than the main nepheloid layer in the deep basin, and at some stations additional layers are present at lesser depths All of these shallower layers are presently considered to be of local extent and to result from processes that occur at basin boundaries, because we have not been able to correlate them with the station spacing that is now available. These intermediate layers are possibly related to the proximity of rivers, such as the Congo River, that supply particulate matter in large quantities and to the interfingering of water masses that occurs on continental slopes. Hydrographic data with a much smaller sampling interval are needed to test the latter hypothesis.

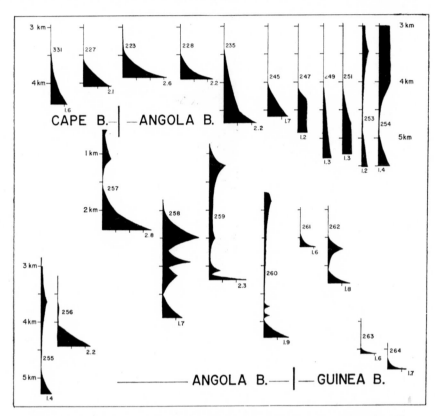

Figure 2 Light scattering profiles in the Cape, Angola, and Guinea Basins. The main feature of each profile is the zone of increased scattering immediately above the bottom. The top of this zone is marked by a minimum in light scattering which is chosen as the reference in the calculation of relative exposure. The steady decrease of light scattering from the sea surface to this minimum is not shown. Distinct layers of increased scattering in the intermediate water above the main nepheloid layer are indicated where present. Positions are given in Figure 1. The nephelometer units are relative film exposure, and the origin of each profile is $E = 1$

In Figures 3, 4, and 5 are presented curves of potential temperature and salinity at stations typical of the Angola Basin. These data indicate homogeneous bottom water below 4500 m with potential temperature 1.95°C and salinity 34.89%. The deviation from these values becomes conspicuous only

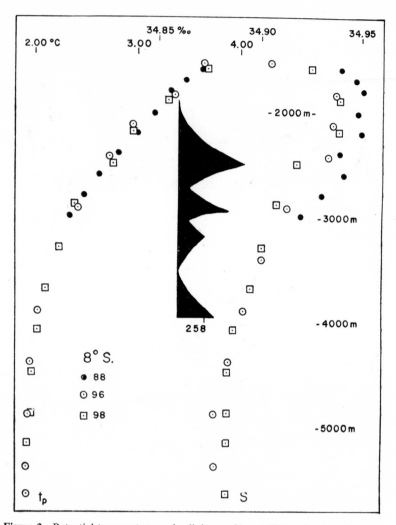

Figure 3 Potential temperature and salinity profiles at 8° S in the Angola Basin. Water below 4000 m is nearly homogeneous and that below about 4700 m has constant potential temperature and salinity. Light scattering zones on the continental slope, e.g. at 258, are not related to the homogeneous water and indicate processes that occur along basin boundaries. Station 258 is close to hydro station 88 and to the mouth of the Congo River. Hydrographic data in this and other figures are from Fuglister (1961), and potential temperatures were calculated according to Wüst (1961)

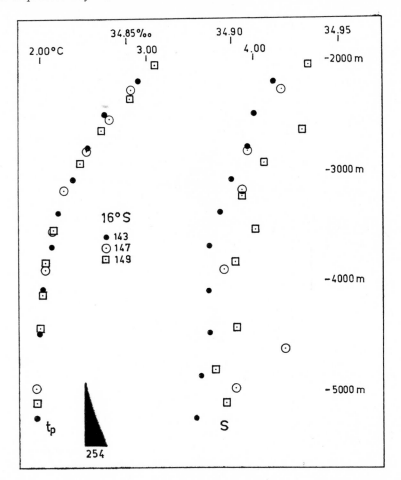

Figure 4 Potential temperature and salinity at 16° S in the Angola Basin. Water below about 4000 m is nearly homogeneous and is homogeneous below 4700 m. The nepheloid layer is associated with this homogeneous zone (compare hydrographic station 147 with nephelometer station 254 (Group 3). Positions are shown in Figure 1. The considerable scatter in the salinity data here is not present at 8° S or 24° S

above 4000 m. The main nepheloid layer is considered to occupy the homogeneous water mass in which the tendency of the particles to settle to the bottom is balanced by turbulence. The presence of a nepheloid layer in homogeneous bottom water has also been observed in the North Pacific (Ewing and Connary, 1971) and in the Gulf of Mexico (Ewing *et al.*, in prep.). The mixing of bottom water creates the zone from 4000 to 4500 m

176

in which there is slight positive stability. This zone marks the upper boundary of the main nepheloid layer.

The dispersal of suspended particles throughout the deep basin is attributed to advection and diffusion in a gyre, the direction or rotation of

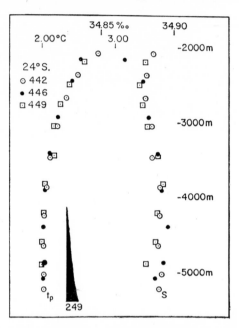

Figure 5 Potential temperature and salinity profiles at 24° S in the Angola Basin. Water below about 3800 m is nearly homogeneous, and is homogeneous below about 4400 m. As at 16° S, the nepheloid layer is found within the homogeneous zone. T-S profiles at this latitude are different from those at 8° S and 16° S in that the homogeneous layer is about 300 m thicker at 24° S. This is attributed to the large-scale stirring effect of high hills shown in Figure 1 between 20 and 26° S. North of this area is predominantly abyssal plain and low relief. Nephelometer station 249 is typical of Group 2, the top of the nepheloid layer being about 800 m shallower than in Group 3 (254, Figure 4) to the northeast

which cannot be inferred with confidence until data from additional suitably located stations become available. In the southern hemisphere a clockwise pattern would be expected from Coriolis force considerations.

III GUINEA BASIN

Four light scattering profiles in the Guinea Basin are shown in Figure 2. A nepheloid layer about 150 m thick is observed in the eastern part of the basin (stations 263, 264). The intensity of scattering is about 1.7, which is

nearly the same as that in the deep Angola Basin. More stations are necessary in this basin before conclusions can be drawn about basin-wide variability of the scattering intensity and of the depth to the top of the nepheloid
layer. At 262 on the African continental slope, a similarly thin bottom layer
is present, but there an additional shallower layer of nearly equal intensity
is present at 2700 m. Presumably the bottom nepheloid layer on the slope
is continuous with that in the deep basin, but a series of closely-spaced
stations from the continental margin to the deep basin is needed to test this
hypothesis.

According to Fuglister (1961), the very cold bottom water in the Romanche Fracture Zone (potential temperature about 0.6 °C) is not found in

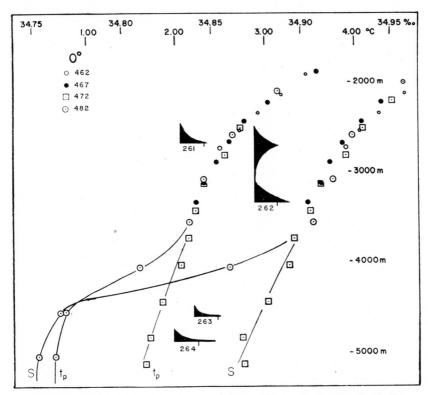

Figure 6 Potential temperature and salinity profiles at the Equator in the Romanche Fracture Zone (482) and Guinea Basin. The bottom water at 5100 m at 472
(Figure 1) has the temperature and salinity of water at about 4000 m at 482. A sill
is suggested between 3700 m and 4000 m. A similar contrast of temperature-
salinity structure exists between 478 and 476, implying that the sill is to be found
between them. Despite the existence of gradients of temperature and salinity, a
nepheloid layer is present on the slope and in the deep basin. The depth scale applies
to both the nepheloid and hydrographic profiles

the Guinea Basin where the bottom potential temperature is about 1.7°C (Figure 6). Water (1.7°C, 34.87‰) at 5100 m in the Guinea Basin (station 472) is, however, found at about 4000 m in the Romanche Fracture Zone (station 482), and a marked divergence in the potential temperature and salinity curves at these two stations occurs at about 3700 m. Allowing for uncertainties in the flow pattern over the sill, these curves suggest a sill at 3700 m to 4000 m between stations 472 and 482. On the basis of this and other data, Metcalf *et al.* (1964) concluded the sill is no deeper than 3750 m.

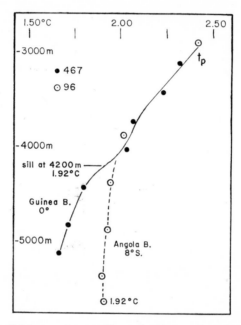

Figure 7 Potential temperature profiles at the Equator in the Guinea Basin and at 8° S in the Angola Basin. The nearly homogeneous water below about 4100 m in the Angola Basin is replenished by water flowing southward from the Guinea Basin over a sill whose sounding is 4100 m to 4200 m

The temperature-salinity structure at stations as far east as station 478 is comparable to that at 482, whereas that eastward of and including 476 is similar to 472. Thus, the effective sill is located between stations 476 and 478 (Figure 1). The sounding at the intermediate station 477 is nearly 4000 m, and its temperature and salinity values are above the depth at which the divergence occurs.

In the Brazil Basin there is a strong nepheloid layer below approximately 4400 m (unpublished data). As only water above 4000 m presumably enters the Guinea Basin and because the concentration of particles in the relatively clear water above the main nepheloid layer is probably small, very little suspended matter enters the Guinea Basin by this route.

Vertical gradients of potential temperature and salinity characterize the bottom water below 4000 m in the Guinea Basin but are virtually absent in the nearly homogeneous bottom water below 4000 m in the Angola Basin. A topographic barrier separating these two basins is suggested, the depth of the sill being at about 4100 m to 4200 m (Figure 7). On the basis of the available nephelometer data, this still is sufficiently high to allow only the relatively clear intermediate water of the Guinea Basin to renew the bottom water of the Angola Basin.

IV WALVIS RIDGE RECONSIDERED

As noted in Section II, Group 2 and 3 nephelometer stations in the deep Angola Basin (16 to 24° S) are roughly the same intensity, whereas Group 1 stations in the southwestern corner of the basin are anomalous in that the main nepheloid layer is significantly stronger. In the region of Groups 2 and 3 the curves of potential temperature and salinity below 2500 m are of similar shape with a slight increase of temperature and decrease in salinity of the homogeneous bottom water toward the south (Figures 8 and 9). At

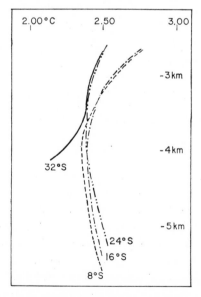

Figure 8 Profiles of *in situ* temperature at 8, 16, 24, and 32° S in the Angola Basin. The cooling of water below 3600 m at 32° S is anomalous. The deep temperature minimum at 3.9 km is a basin-wide feature. No such minimum is observed in the Guinea or Cape Basin

32° S, in the area of Group 1, different vertical profiles are seen, indicating a different kind of bottom water, whose temperature and salinity decrease with depth.

The nepheloid layer in the deep part of Cape Basin is strong with E values in the range of 3.0 to 6.0. The existence of a deep gap in the Walvis Ridge through which particle-laden water flows from the Cape Basin into the southwestern corner of the Angola Basin could explain the stronger

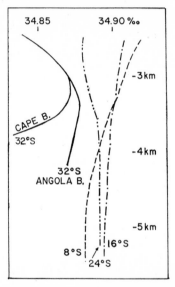

Figure 9 Profiles of salinity at 8°, 16°, 24°, and 32° in the Angola and Cape Basins. From 8° to 24° S salinity below the temperature minimum is essentially uniform. At 32° S in the Angola Basin, the decrease of salinity with depth below 3600 m is anomalous, but is a pattern more similar to the salinity structure in the Cape Basin at 5832

nepheloid layer and anomalous temperature and salinity structure in the Group 1 area. The similarity in the level of the clearest water on the north and south flanks of the Walvis Ridge (compare stations 331 and 227 with 223 and 228, Figure 2) would be also explained. A comparison of *in situ* temperature profiles (Figure 10) shows the similarity in temperature structure between the Cape Basin and the Group 1 area and the difference between these and the Group 2 region at 24° S. Salinity profiles (Figure 9) in the Angola Basin (32° S) are of essentially similar shape as those in the Cape Basin, although the decrease in salinity with depth below 3300 m is not large in the Angola Basin (32° S). Soundings obtained from recent cruises of *Vema* and *Robert D. Conrad* show that the Walvis Ridge is not conti-

nuous and at its southwestern end may be defined by a number of discon-
nected topographic highs, guyot-like features, with narrow deep channels
separating them.

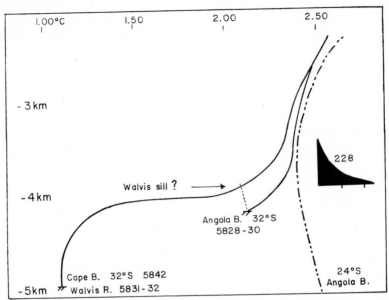

Figure 10 *In situ* temperature profiles in the Cape and Angola Basins at 32° S.
The southwestern Angola Basin (32° S) is hydrographically more similar to the
Cape Basin than to the Angola Basin at 24° S. A sill depth of at least 3900 m
is required to account for the cold bottom water at 5828–30 in the southwestern
Angola Basin. The nepheloid layer at 32° S is associated with the non-homo-
geneous water in the Angola (228) and Cape (331) Basins. The top of the layer
is at about the same depth at these Group 1 stations on opposite sides of the
Walvis Ridge (Figure 2)

At present the explanation for an abrupt rather than a gradual transition
of bottom water properties between the Group 1 and Groups 2 and 3 areas
is unresolved. No clearly defined ridge is evident from available topographic
data in the region between nephelometer stations 235 and 245. Several
crossings by *Vema* and *Robert D. Conrad* show that the region between
20° S and 26° S (Figure 1) is an area of rough topography. The depth to
some peaks is about 4400 m, and some have a relief of as much as 1000 m.
Whether any of these peaks forms a continuous barrier cannot yet be inferred. Seismic reflection profiles through this area indicate a scarcity of
sediments in many of the numerous troughs; highly stratified sediments
indicative of turbidites floor some valleys. More crossings are needed to
ascertain the extent to which the troughs may be interconnected and how

they are related to the Angola Abyssal Plain. It is our belief that this region of high relief prevents significant mixing of central and southwestern Angola Basin nepheloid water. Some continuous passages through these hills may exist, but it is thought that they are small in number and size and provide a means for only limited exchange.

Although the scattering intensity of Group 2 and 3 stations is the same, the two groups may be distinguished by a difference in depth to the top of the nepheloid layer (Figure 2). This difference cannot be attributed to a difference in depth of water as soundings are nearly the same at these stations. Group 2 stations are located within the area of rough topography outlined in Figure 1, and Group 3 stations are to north on the smooth Angola Abyssal Plain. Hills of high relief compared to smooth abyssal plain probably have a greater stirring effect on a nearly homogeneous water column, and in such an area, suspended particles can be carried to higher levels.

V SUMMARY

The bottom water in the Angola Basin has been thought to be renewed by a flow entering the basin at its northern margin. Recent hydrographic data support this long-standing conclusion. Temperature and salinity profiles in the southwestern corner (32° S) of the Angola Basin do not, however, fit this scheme of bottom circulation. Based on this data, the most reasonable explanation for the anomaly is the presence of gaps in the Walvis Ridge through which bottom water of the Cape Basin passes into the southwestern Angola Basin. The sill depth of these gaps is not certain from the data analysed, but it appears that one or more sills deeper than 3900 m is required. Unpublished topographic data indicate that the sill is probably not much deeper than 4000 m. The contrast of temperature and salinity profiles at 32° S with those from 8°, 16°, and 24° S seems most reasonably explained by an intervening barrier which prevents the influxing Cape Basin water from spreading northward into the main Angola Basin. A large region of rough topography between about 20° to 26° S may be such a barrier. Sills of 4000 m are present in the vicinity of the Romanche Fracture Zone and divide the Guinea and Angola Basins. These sills determine the structure of the water column below sill depth.

Light scattering intensities in the bottom waters of the Angola and Guinea Basins are weaker than those in the northwest and southwest Atlantic by a factor of about 5. They are roughly equal to those in the north and central Pacific.

The generally intense nepheloid layer in the Cape Basin (a factor of 3 to

6 increase in scattering relative to clear intermediate water) and the weak layer of Angola Basin (factor of about 1.4) suggest that the bottom circulation in each might be unrelated. Moderately strong light scattering intensities in the southwestern Angola Basin north of the Walvis Ridge appear anomalous and support the claim that there is some bottom water communication between these basins. The strength of the layer, except in this anomalous region, is directly related to the proximity to the African continent, which is postulated to be the major source of the particulate matter in the nepheloid layer. Few particles are carried into the Angola Basin by deep water from the Guinea Basin. On the Angola Basin continental slope, one or more shallower scattering zones are common above the main nepheloid layer.

Acknowledgement

Drs. Arnold Gordon and Kenneth Hunkins read the manuscript and gave helpful comments. This research was sponsored by The Department of the Navy, Office of Naval Research, Contract N00014-67-A-0108-0004.

References

Eittreim, S.L., Suspended matter in the deep waters of the northwest Atlantic Ocean. Ph.D. Dissertation, Columbia University, 152 p., 1970.

Eittreim, S. and M.Ewing, Suspended particulate matter in the deep waters of the North American Basin, (this volume) 1971.

Eittreim, S., M.Ewing, and E.M.Thorndike, Suspended matter along the continental margin of the North American Basin. *Deep-Sea Research*, 16, p. 613–624, 1969.

Eittreim, S., A.L.Gordon, E.M.Thorndike, and P.Bruchhausen, The nepheloid layer and observed bottom currents in the Indian-Pacific Antarctic Sea, (this volume) 1971.

Ewing, M. and S.D.Connary, Nepheloid layer in the North Pacific. *Memoir Geol. Soc. Amer.*, 126, J.Hays, editor, GSA 1971.

Ewing, M., S.Eittreim, J.Ewing, and X. Le Pichon, Sediment transport and distribution in the Argentine Basin: Part 3, Nepheloid layer and processes of sedimentation. *Physics and Chemistry of the Earth*, S.K.Runcorn, editor, 8, Pergamon Press Ltd., Oxford, 1971.

Ewing, M., E.M.Thorndike, and L.G.Sullivan, Nepheloid layer in the Gulf of Mexico, in prep.

Fuglister, F.C., Atlantic Ocean atlas of temperature and salinity profiles and data from the *International Geophysical Year 1957–58. Atlas Ser. Woods Hole Oceanogr. Inst.*, 1, 1961.

Hunkins, K., E.M.Thorndike, and G.Mathieu, Nepheloid layers and bottom currents in the Arctic Ocean. *J. Geophys. Res.*, 74, 6995–7008, 1970.

Jacobs, M., E.M.Thorndike, and M.Ewing, Light scatterers of the deep sea: the nature of suspended particulate matter, in preparation.

Metcalf, W.G., B.C.Heezen, and M.C.Stalcup, The sill depth of the Mid-Atlantic Ridge in the equatorial region. *Deep-Sea Research*, 11, pp. 1–10, 1964.

Thorndike, E.M. and M. Ewing, Photographic nephelometers for the deep sea. In: Deep Sea Photography, J.B. Hersey, editor. Johns Hopkins Press, p. 113–116, 1967.

Wüst, G., Das Bodenwasser und die Gliederung der Atlantischen Tiefsee. *Wiss. Ergeb. dt. atl. Exped. Meteor* 1925–1927, **6**, (1) (1), p. 1–107, 1933.

Wüst, G., Stromgeschwindigkeiten und Strommengen in den Tiefen des Atlantischen Ozeans. *Wiss. Ergeb. Dt. Atl. Exp. Meteor* 1925–1927, **6**, (2) (6), pp. 35–420, 1957.

Wüst, G., Tables for rapid computation of potential temperatures. Technical Report CM-9-61-At (30-1) 1808. Geol. of Columbia University, New York, 1961.

Sources of Mediterranean Intermediate Water in the Levantine Sea

SELIM A. MORCOS

Oceanography Department, Faculty of Science
University of Alexandria, Egypt, U.A.R.

Abstract   The surface waters of the northern Levant Sea (eastern Mediterrean) are slightly cooler than that to the south. Surface salinity is higher than in the south during the summer, but similar during winter, when the southern region attains its maximum value. Hydrographic stations in the two regions suggest winter time vertical mixing. The four seasonal cruises of *"Ichthyolog"* 1966 near the Egyptian delta and Sinai show a homogeneous dense surface water in February, which is transformed into the well developed intermediate salinity maximum in August and November. *"Shoyo-Maru"* section along the Egyptian coast in March 1959, indicates a homogeneous saline upper layer in the east which spreads as intermediate layer into the western part of the section. A secondary source of formation of intermediate water is detected in the southern Levant. Intermediate water in the Levant is more heterogeneous than in the other basins of the Mediterranean, which indicates various source regions with respect to space and time. These waters by mixing acquire more homogeneity on spreading towards the Ionian Sea, and is recognized as a well identified water mass west of the Strait of Sicily.

INTRODUCTION

The presence of an intermediate water characterised by a secondary maximum salinity in the Mediterranean is a very peculiar phenomenon in this sea. No such phenomenon is present in a similarly semiclosed sea in an arid zone like the Red Sea. Nielsen (1912) attributed this layer to an intermediate outflow from the Levant Sea since the distribution of salinities in the intermediate layer shows a maximum which gradually increases towards the Levant. From the study of two *"Thor"* Stations near Crete and Rhodes, he concluded that "in the central and northern parts of the Levant, where there is no intermediate minimum of salinity, and where the surface water is highly saline (over 39‰), a homohaline surface layer is formed in winter.

185

As the cooling process is not so strong that this layer can sink down into the depths, it spreads out towards the west as an intermediate current."

Wüst (1960) pointed out that along the coasts of Asia Minor, the temperature drops in February to values between 12.5° and 15.5 °C. At the same time the surface salinity reaches its maximum of about 39.1 $^o/_{oo}$. As a result of the influence of low temperature and high salinity, a relatively dense surface water is formed on both sides of Rhodes. This homogeneous large water mass is formed in the upper 250 m, from where it spreads in the core layer of what Wüst has called the Levantine Intermediate Water. While spreading westward, it mixes with lower salinity water from above and below, resulting in a decrease in salinity and also temperature, with an associated increase in depth (50–100 m to 100–250 m and finally to 300–400 m). Wüst used for this investigation 182 winter stations and 341 summer stations from all the basins of the Mediterranean during the period 1908 to 1958. Only 27 winter and 49 summer stations were available from the largest basin, the Levantine. Part of the difficulty in studying this problem is the lack of data, especially in winter.

There is a general agreement among oceanographers that the northern part of the Levant is the source region of the intermediate water (Nielsen, 1912; Wüst, 1959, 1960 and 1961; Lacombe and Tchernia, 1960; Miller, 1963; Ovchinnikov and Fedoseyev, 1965; Moskalenko and Ovchinnikov, 1965; Oren and Engel, 1965 and Bogdanova and Lebedeva, unpublished).

The author has previously expressed the view that the southern Levant may be a possible source of the intermediate water (Morcos, 1967). The aim of this paper is to examine critically this view by comparing the climatological and hydrographic factors in the northern and southern Levant, and by studying the new data collected during the last decade from the southern Levant which is one of the least investigated areas in the Mediterranean. The stations available from the southern Levant (south of 33° N) during 1908–1958 were 9 winter stations and 8 summer stations. Only few stations were added to the archive of the National Oceanographic Data Center (Washington, D.C.) for this region during the last decade. The present investigation will make use of about 100 deep stations collected during joint programs with Egyptian oceancgraphers on board the following research vessels: the Japanese "*Shoyo-Maru*" in March 1959 (Gorgy and Shaheen, 1964), the Yugoslavian "*Ovicica*" and "*Globica*" in October 1959–1961, the USSR "*Ichthyolog*" in October 1964 (Halim *et al.*, 1967) and December 1965 to December 1966 (Hassan, 1969; Emara, 1969). These observations improve our understanding of the oceanographic conditions in the southern Levant. Part of these observations will be used in the present paper together with recent observations from the rest of the sea.

Of particular interest are Wüst's T-S diagrams for the core layer of the Levantine Intermediate Water in winter and summer (Figure 1). All points representing the core layer aggregate along the isopycnal of $\sigma_t = 29\,05$, except those of the Levant which spread between σ_t 24.1 and 28.6. Such scatter of the Levantine data cannot be explained by the fact that the diagrams depend on data from 11 ships over 50 years, because the data representing the Levantine are almost exclusively obtained from R/V "*Atlantis*" in April 1948, and R/V "*Calypso*" in October 1955. This may cast some doubt on the assumption that the source of the intermediate layer is one homogeneous water mass in the north of the Levantine basin. We possess no direct measurements or observations from this region during the coldest time of the year similar to the attempt done by three research ships in winter 1969 to observe the formation and spreading of the deep and bottom water

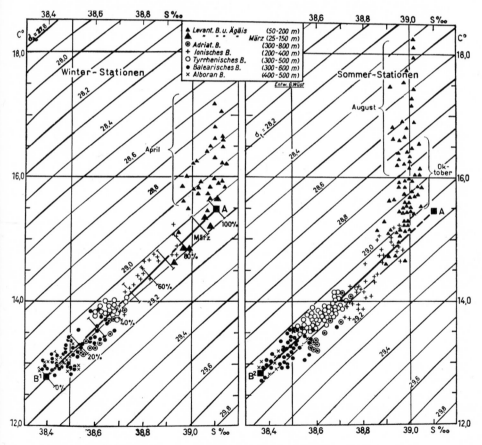

Figure 1   T-S diagram for the core layer of the Levantine Intermediate water in all basins of the Mediterranean Sea (After Wüst, 1960)

in the northern border regions of the western Mediterranean (Tchernia, personal communication, 1969). The best time for observing such phenomenon would be February–March after the minimum air temperature occurs in the Mediterranean. The most suitable stations available to the Wüst (1960) and Lacombe and Tchernia (1960) investigations were the "*Atlantis*" stations in April 1948. At that time, warming started to develop in the upper layer.

CLIMATOLOGICAL AND HYDROGRAPHIC FACTORS IN THE NORTH AND SOUTH LEVANT

The formation of intermediate water depends mainly on climatological conditions resulting in vertical mixing and sinking of relatively cold saline waters to intermediate depths. A brief comparison between climatological conditions in the northern and southern regions follows.

Table 1 Air Temperature in Coastal Meteorological Stations in the coldest month of the year (Jan. or Feb.) in North and South Levant Sea

(From Weather in the Mediterranean, Vol. II, 1962)

Air Temperature	Northern Levant Sea (°C)			Southern Levant Sea (°C)		
	Rhodes	Antalya	Cyprus Cape Andreas	Port Said	Alex- andria	Saloum
Daily Mean	11.7	10.6	12.2	13.5	13.9	12.2
Mean of Daily Max.	15.6	14.4	15.6	18.9	18.3	17.2
Mean of Daily Min.	7.8	6.1	8.9	10.6	10.6	7.2
Daily Range	7.8	8.3	6.7	8.3	7.8	10.0
Mean of Monthly Max.	19.4	18.3	19.4	23.3	22.8	22.8
Mean of Monthly Min.	2.8	0.6	2.8	7.2	6.7	3.9
Monthly range	16.6	17.7	16.6	16.1	16.1	18.9

Table 1 compares between the air temperature in some meteorological stations in the northern and southern Levant in the coolest month of the year (January or February). With the exception of Antalya which reflects continental characteristics due to its deeply situated position in a bay on the Turkish coast, the daily minimum in the north is about 1.2 °C less than in the south, the monthly minimum is 3.1 °C less. The range between the daily maximum and minimum, which may influence the diurnal variation of sea water temperature and which may initiate sinking of water in cold nights, is however, greater in the south by about 1.3°C. Comparison is made between the two large islands Rhodes and Cyprus and the flat exposed Egyptian coast.

Table 2 Monthly average temperature of sea surface and overlying air in a strip of 5° squares along the Turkish and Egyptian Coasts (°C)

Months	Turkish Coast (°C)			Egyptian Coast (°C)		
	Sea Water	Air	$t_w - t_a$	Sea Water	Air	$t_w - t_a$
January	17.0	15.2	1.8	17.2	15.9	1.3
February	15.5+	14.4+	1.1	16.3	15.4+	0.9
March	16.2	15.6	0.6	16.0+	16.3	−0.3
April	16.9	17.3	−0.4	17.4	17.9	−0.5
May	19.5	20.2	−0.7	20.0	20.6	−0.6
June	22.0	23.5	−1.5	22.9	23.5	−0.6
July	24.8	25.8	−1.0	24.9	25.5	−0.6
August	26.5	27.1	−0.6	25.8	26.2	−0.4
September	25.2	25.4	−0.2	25.4	25.2	0.2
October	23.3	23.1	0.2	24.0	23.5	0.5
November	20.6	19.6	1.0	21.9	20.8	1.1
December	18.2	16.6	1.6	19.2	17.8	1.4
Annual range	11.0	12.7		9.8	10.8	

The differences between the north and south will be obviously less when considering maritime air over the sea surface and much more than the surface water temperature itself. We are indebted to the monthly charts of the Netherlands Meteorological Atlas (1957) which give the average values of air temperature, sea surface temperature (together with the number of observations) for every one-degree square in the Mediterranean. Table 2 gives the average values of air and surface temperatures calculated for a strip of five squares in front of the Turkish and Egyptian coasts. The table shows that both the northern waters and overlying air are only cooler by 0.5°C on the average than in the south, except in the three summer months when the air over the Turkish waters is warmer than over the Egyptian waters, which have also relatively low temperature in August. It seems that the occasional cold and dry subpolar continental air masses have only a limited and a temporary effect on the coastal Turkish waters. "Both summer and winter temperatures on the Karamanian coast are greatly affected by local orography. In summer hot föhn winds may descend from Taurus Mountains (6500 feet), in winter, on the other hand, there are very cold bora winds. The ordinary action of the mountains, however, is to protect the coasts from cold N. winds and to strengthen the sea breeze in summer" (Weather in the Mediterranean, Vol. II, 1936)

Table 3   Average monthly precipitation in coastal Meteorological Station in the Northern and Southern Levant (mm)

(From Weather in the Mediterranean, Vol. II, 1962)

Month	Northern Levant Sea (mm)			Southern Levant Sea (mm)		
	Rhodes	Antalya	Cyprus Cape Andreas	Port Said	Alex-andria	Saloum
January	*237*	247	96	*18*	48	21
February	123	151	74	13	24	15
March	110	75	47	10	11	8
April	25	39	36	5	3	<1
May	26	27	22	3	2	5
June	0.5	13	2	0.0	0.0	<1
July	0.0	2	0.0	0.0	0.0	<1
August	0.0	1	<0.1	0.0	<0.1	0.0
September	4	11	16	0.0	1	<1
October	96	49	26	3	6	2
November	152	129	61	10	33	*26*
December	199	*267*	*138*	15	*56*	18
Total	973	1011	518	77	184	95

One of the factors previously considered is the influence of the continental dry air which enhances evaporation and formation of intermediate water in the northern Levant sea. However, Table 3 shows that the most rainy region in the Levantine is the south coast of Asia Minor with about 900 to 1000 mm annual rainfall and very low rainfall—(less than 100 mm/ year) are received by the Egyptian coast" (Mittelmeer Handbuch, V. Teil, Die Levante, 1965).

SURFACE SALINITY

According to Wüst (1960) the seasonal variation of the surface salinity in the Mediterranean is affected primarily by evaporation and to a secondary extent by the annual cycle of rain fall. Table 2 shows that in both northern and southern Levant sea, the water temperature is higher than air temperature in fall and winter by an average of 1 °C, a condition which favours evaporation and cooling. Evaporation in the Northern Hemisphere and the Mediterranean Sea is at minimum in summer, May–July, and at a maximum in fall and early winter, October–December (Sverdrup *et al.*, 1942; Wüst, 1959b). The annual cycle of salinity is partly affected by the annual cycle of rain (see Table 3), and the salinity maximum in the Mediterranean is shifted to the end of summer, September–November, and the salinity minimum to the end of winter, April–June (Wüst, 1960).

We possess now various observations of the seasonal variation of salinity of the water of the Egyptian coast. Before the erection of Aswan High Dam, a low minimum was always observed in September and a maximum in February–April in Port-Said, and in November–December in Alexandria (Morcos, 1960; El-Maghraby and Halim, 1965). During the last normal Nile flood (1964), Halim *et al.* (1967) have observed a thin layer of low salinity water spreading over a very dense layer (greater than $39.40^{\circ}/_{oo}$ salinity), which presumably appears at surface after the end of the flood in November. Monthly observations at a station on the continental shelf north of Alexandria, have demonstrated a minimum of $38.6^{\circ}/_{oo}$ in April 1964, a period of low salinity (38.70 to $38.80^{\circ}/_{oo}$) from May to August and a period of high salinity ($>38.9^{\circ}/_{oo}$) from October to February with a maximum of $39.16^{\circ}/_{oo}$ in November 1964 (El-Kirsh, 1969). The four seasonal cruises of the "*Ichthyolog*" in 1966 off the Egyptian Delta and Sinai indicated a minimum salinity in April (38.9–$39.0^{\circ}/_{oo}$) and a maximum in February (39.3–$39.4^{\circ}/_{oo}$) (Hassan, 1969).

The surface salinity in the northern Levant Sea in February–March is $38.91^{\circ}/_{oo}$ (average of 13 stations of "*Atlantis*" 1962 and "*Chypre-04*" 1965). This value is lower than that in February 1966 as found by the "*Ichthyolog*", (39.3–$39.4^{\circ}/_{oo}$) but very close to $38.95^{\circ}/_{oo}$ the average of three "*Atlantis* II" stations in the southern Levant in February 1965. The surface salinity at the time of formation of the intermediate water in winter, is similar at the northern and southern parts of the Levant, if not slightly higher in the south.

A totally different picture is obtained from a published chart for surface salinity in winter, which shows values higher than $39.1^{\circ}/_{oo}$ around Rhodes and 38.4–$38.8^{\circ}/_{oo}$ in front of the Egyptian coast (Wüst, 1960; Lacombe and Tchernia, 1960). The only winter stations available from the Levant at that time were those of *Atlantis* 1948, and this chart depends on stations taken in the first week of April around Rhodes and in the following fortnight opposite to the Egyptian coast. This is the time when the minimum salinity occurs in the Egyptian waters, after the very distinct maximum of the preceeding winter. It seems that the minimum salinity occurs earlier in the southern Levant, presumably under the influence of the Atlantic Ocean surface water entering from the west. This current, characterised by a sub-surface minimum salinity, is well developed only during the summer at 20 to 75 m depth, along the North African continental slope as clearly shown by the core chart of Lacombe and Tchernia (1960, Figure 3) and Moskalenko and Ovchinnikov (1965, Figure 5).

Our knowledge of the seasonal and regional distribution of salinity in the Levant is still not sufficient and further observations are needed. Comparison between data from different years may lead to some discrepancy. Changes in salinity from year to year have been observed by Sverdrup *et al.* (1942) in the Ionian Sea, and by Miller (1963) in the north eastern Levantine Sea. One is curious about the effects of Aswan High Dam not only on the circulation near Port Said and the delta area but also of the circulation of the entire eastern Mediterranean and the production of the Levantine intermediate water. Perhaps the intermediate water would be composed of slightly higher salinities and thus occur at deeper depths than previous years (Arnold L. Gordon, personal communication, 1967).

HYDROGRAPHIC STATIONS IN THE NORTH AND SOUTH LEVANT

A net work of hydrographic stations were carried out by R/V *"Ichthyolog"* in the southern Levant (south of 33° N) along the Egyptian coast. The same stations were repeated in February, April, August and November 1966. They constitute six meridional sections distributed between 29° and 34° E. Each section extends from the coast to the deepest station (about 50 miles offshore). The northernmost station in the westernmost section is chosen to represent the conditions off the Egyptian coast. This section runs parallel and to the east of long. 29° E (and west of Alexandria). The area of this station lies in Marsden Square 142 subsection (19), and will be designated in this discussion as area A. For the sake of comparison two other areas are picked in the vicinity of Rhodes, where, according to Wüst (1960), the most favourable conditions for the formation of the Intermediate Water prevail. The two areas, designated B_1 and B_2 are situated in Marsden Squares 142-(68) and 142-(58), to the north east and south east of Rhodes respectively. The two areas are well represented by stations from *"Atlantis"* in March 1962, *"Akademik S. Vavilov"* in September 1959, and *"Chain"* in November 1961. Area B_1 is enclosed between Rhodes Island and the Turkish coast and has a depth of about 650–950 m, while B_2 has a depth of 3500 to 4500 m. The station representing area A lies on the continental slope and has a depth of > 500 to 1500 m due to small changes in its position in the four seasonal cruises.

Figure 2 compares the vertical distribution of temperature, salinity, σ_t, and oxygen in the four seasons in Area A with those of area B_1 (March and September) and area B_2 (March and November).

In winter (February–March) the vertical gradient of temperature is very weak or completely absent, and a mixed layer extends down to various depths. In the shallower northern area B_1, a nearly isothermal layer occupies the upper 250 m where the surface temperature is only $0.07\,^\circ$C lower than area A temperatures at 250 m. In areas B_2 and A, a slight decrease in temperature with depth is observed in the upper 100 m. A perfectly isohaline layer of $39.27\,^\circ/_{00}$ extends down to 300 m in area A. The salinity in area B_1 shows small deviations from $38.83\,^\circ/_{00}$ in the upper 250 m, while it exhibits a very weak intermediate maximum between 50 and 150 m in area B_2. The density decreases slightly with depth in the upper 250 m in area B_1, thus indicating neutral or perhaps unstable stratification. In areas A and B_2, a very small gradient occurs due to surface heating. Further evidence for vertical mixing appears in the three curves of oxygen. A well mixed layer appears in the upper 250 m in area B_1, while it is restricted to the upper 100 m in A and B_2.

In spring (April), the upper layer in area A becomes warmer, less saline, and less dense. Relatively stronger gradients start to develop in the three curves at some intermediate depth.

In summer (August–September), three layers are distinguished in the temperature curves, a hot homogeneous layer due to mixing by wind in the upper 25 m, followed by a strong thermocline which relaxes below 150 m. Two inversions appear in the salinity distribution, a subsurface minimum at about 50–125 m, and an intermediate maximum at about 200–250 m. The strongest vertical gradient of density appears in this season due to the very strong thermocline. A subsurface maximum oxygen occurs in both stations at nearly the same level of subsurface minimum salinity.

In the fall (November), the curves are generally similar to those of summer. The upper mixed layer becomes deeper due to the effect of wind and beginning of cooling after October.

In a similar study in the southern Aegean, Bruce and Charnock (1965) noted that winter cooling, having relaxed the strong thermocline in the preceding fall, allows the high salinity layer produced during the summer to sink. In a similar manner the summer salinity submaximum also must sink slightly. Thus a source of cold, high salinity water exists during the winter which seems to sink to the depth of spring intermediate maximum.

Figure 2 exhibits the great similarity of the main hydrographic features

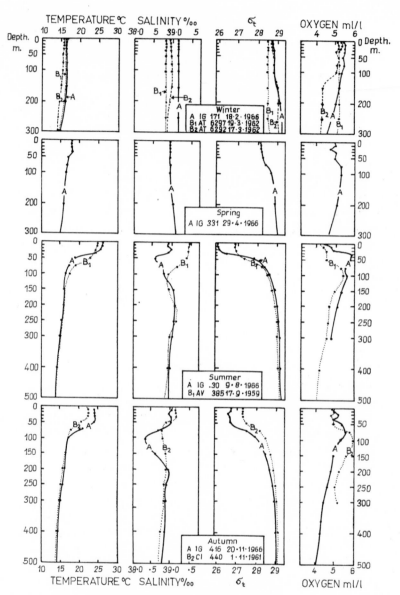

Figure 2 Vertical distribution of temperature, salinity, density and oxygen, in the four seasons, in hydrographic stations in area A north of Alexandria, area B_1 north east of Rhodes, and in area B_2 south east of Rhodes. Research Vessels are "*Ichthyolog*" (IG), "*Atlantis*" (AT), "*Akademik S. Vavilov*" (AV), and "*Chain*" (CI)

in the south and north of the Levantine Basin. However, some differences can be detected. In winter, the water column in the south has similar temperature or slightly higher than found in the north. It is more saline, and as a result its density is of the same magnitude if not higher than some of the stations in the north. In summer, and autumn, the water columns in the south and north show close values of temperature, salinity and density below 150–200 m depth. The layer of subsurface minimum salinity is better developed in the south. The upper layer is warmer, less saline, and less dense than in the north. This may exhibit opposite trends in the seasonal

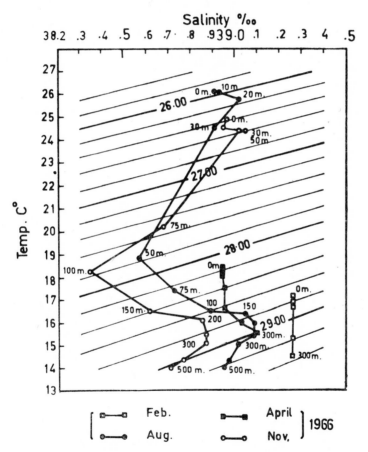

Figure 3 T-S of "*Ichthyolog*" stations in the four seasons in area A in the Southern Levant.

Winter   IG 171   29°04′ E, 31°39′ N
Spring   IG 331   29°04′ E, 31°44′ N
Summer   IG 30   20°05′ E, 31°40′ N
Autumn IG 416   29°05′ E, 31°20′ N

variation of salinity. The salinity shows a maximum in winter in the south, while in the north highest values are attained in summer and autumn.

Figure 3 represents the T-S diagrams of the station representing area *A* near the Egyptian coast in the four seasons during 1966. In February a completely homogeneous layer occupies the upper 300 m, and is characterised by a small range of temperature, and the annual maximum of salinity (39.27 $^\circ/_{oo}$) and density (28.8–29.4 σ_t). In April distinct stratification occurs in the water column with the upper 100 m becoming less saline and warmer. In August and November, three layers are distinguished, the surface warm layer having salinity less than February but higher than April, the subsurface minimum salinity (50–100 m), and the intermediate maximum salinity (150–300 m). The figure shows that the upper homogeneous salinity layer in April is replaced by a subsurface minimum salinity layer with higher surface salinity values in summer either due to excessive evaporation and heating which result in high salinity and low density at surface, or due to the effect of low salinity water of Atlantic origin flowing from the west, or both factors. The deeper more saline layer (100–300 m) in April will become the intermediate layer of maximum salinity in September and November.

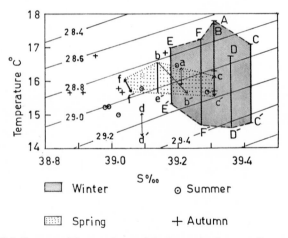

Figure 4   T-S diagram of the core layer of the Levantine Intermediate water in the six northernmost stations of the meridional sections of *"Ichthyolog"* in the four seasons of 1966

It is not difficult to conclude that this seasonal change in the water column indicates a process of formation of intermediate water similar to that suggested by Lacombe and Tchernia (1960) for the Rhodes–Cyprus region where a homohaline layer of 39.1 $^\circ/_{oo}$ salinity occupies the upper 100 m in winter. During the following summer, the base of this layer will not be affected by the sea–air exchange due to the presence of low density surface layer resulting from heating and the presence of the Atlantic effect at about 30–50 m depth. Accordingly, at the end of summer, the base of the homo-haline layer which remains from winter is revealed by a maximum salinity which is encountered in summer at about 100–150 m at the east and 250–300 m at the west of the eastern basin of the Mediterranean.

The T-S diagram in Figure 4 illustrates this seasonal transformation in the northermost stations of the six meridional sections of *"Ichthyolog"* 1966 (Sections A to F from east to west respectively). In winter the six sta-tions show perfect homohaline conditions from 0 to 300 m and are shown on the diagram as vertical lines *A-A'* and *F-F'*. The saline water mass of winter is represented by the space enclosed between *CC'* and *EE'*. In spring, the base of the water column, usually between 200 and 300 m, repre-sents a homohaline layer of intermediate maximum salinity. It is less saline, and has a smaller range of temperature than the larger water mass of winter. Also, it becomes more isolated from the atmosphere by the overlying 200 m layer. In August and November, the core layer at the six stations is repre-sented by points which almost retain the same density but become cooler and less saline than in spring. Although the core layer in these two seasons lies mostly at 150–250 m, its temperature is less or at least equal to the tem-perature at 300 m in spring, which is represented in the diagram by the line *c'f'*. (The temperatures of waters at 300 m in August and November are less than in April.) Compared with the Turkish coast, the continental shelf in front of the Egyptian Delta and Sinai is exceptionally broad. Our observations in the last few years indicate that the water on the continental shelf is warmer in summer and cooler in winter than in the offshore waters (El-Kirsh, 1969; Hassan, 1969). This may form a good source of heavy water in winter away from local dilution from the Nile and Delta Lakes which is usually minimal in this season. The network of *"Ichthyolog"* sta-tions in winter show a very homogeneous water mass of low temperature, high salinity and high density. This dense water mass occupies the broad continental shelf and part of the continental slope which creates favourable conditions for this water to slope down and spread as an intermediate water to the north and west, as has been demonstrated by charts for the distribu-tion of salinity in the core of the intermediate layer of maximum salinity in the four seasons in the southern Levant (unpublished, Hassan, 1969).

One of the most interesting set of observations which represent winter con-ditions in the southern Levant Sea, is the hydrographic section of *"Shoyo-Maru"* in March 1959. This section (Figures 5a, b, c and d) runs along the Egyptian coast just north of Latitude 32° N, between 26° E and 32° E. The temperature section indicates a weak gradient in the upper 100 m. The

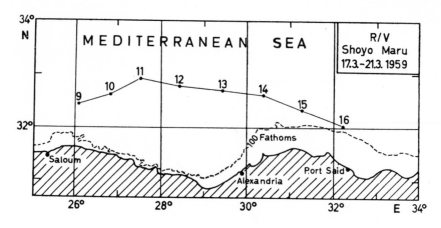

5a   Position of stations

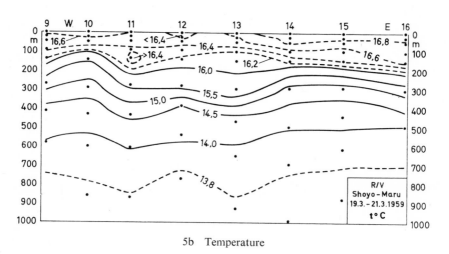

5b   Temperature

temperature of the upper layer is slightly less than area A (near St. 13 in the section) in February 1966, but lies in the range of temperature encountered in area B_1 and B_2 near Rhodes in March 1962. There is evidence that winter 1959 was particularly severe (Morcos, 1970).

A layer of relatively high salinity of 38.9 to 39.1 °/₀₀ occupies the upper 300–400 m in the eastern flank of the section, from where it extends as a layer of intermediate maximum salinity between 100 and 300 m in the western part of the section. The salinity and thickness of this water mass decrease in a westward direction. In contrast to the eastern part of the section, the upper layer of the western stations is occupied by a water mass of low

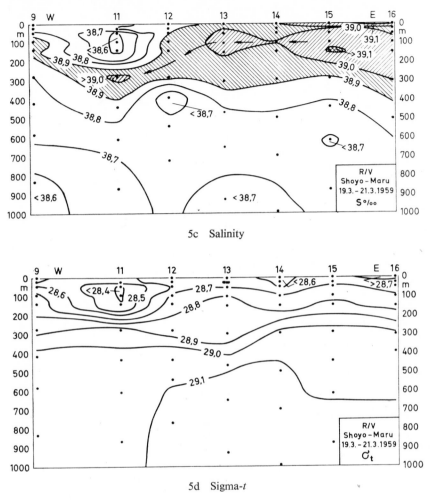

5c   Salinity

5d   Sigma-*t*

Figure 5 *"Shoyo-Maru"* section in the southern Levant Sea

salinity. They even have a subsurface minimum which is very obvious at the northernmost station 11.

At the station (11) ,the developed subsurface minimum is a manifestation of the flow of water of Atlantic origin which occurs mostly to the north of the section. Since the upper stratum of the intermediate layer has been eroded by the Atlantic current, the intermediate maximum appeared at the deeper and colder level at station 11.

The eastern stations (13 to 16) show great similarity to the conditions of February 1966 in area A when the upper 300 m layer is occupied by a highly saline cold water. The western stations (9 to 12) resemble the condition in April in area A with respect to salinity only. Although the upper layer is occupied by low salinity water, yet it still retains the low temperature of winter.

The salinity section can be interpreted in two different ways. One assumption is that the upper layer in the whole section was occupied by highly saline water in the preceding winter and that stratification starts to develop in the west, and at the time of making this section, is in progress towards the east. An alternative interpretation takes in consideration that mixing by winter convection is hindered in the west by the well developed subsurface minimum layer which is maintained by waters of Atlantic origin from the west. The author is inclined to accept the second assumption which is confirmed by *"Ichthyolog"* observations in the eastern region in 1966, and which suggests that this region may play a role in the formation of intermediate water.

Table 4   The core of the layer of the Intermediate
Maximum Salinity in *"Shoyo-Maru"* Section

Station	9	11	12[1]	13[1]	14	15	16
Depth of core layer (m)	91	276	116	124	98	144	150
Int. Max. $S\,^o/_{oo}$	38.99	39.03	38.96	39.02	39.01	39.12	39.07
Temp. at the core (°C)	16.43	15.64	16.26	16.13	16.54	16.44	16.64
Surface $S\,^o/_{oo}$	38.71	(38.80)[2]	38.83	38.95	38.96	39.06	39.16

[1]) The average of two successive depths when the salinity values are identical or show a difference of 0.01∞.
[2]) Salinity at 23 m.

Table 4 gives the depth, salinity and temperature at the core of the layer of intermediate maximum salinity, together with the surface salinity. These data are plotted in a T-S diagram (Figure 6). With the exception of station 11, the points representing the core layer at the different stations vary within $0.15\,^o/_{oo}$ salinity, 0.5 °C temperature, and $0.1\,\sigma_t$. Most of the stations are distributed along the σ_t line 28.74 which indicates lateral mixing or nearly isentropic advection. When this line is extended to show higher temperature and salinity, it meets the point (16 S.W) which represents the surface water of station 16. This may suggest some sort of oblique advective processes between the surface water of the most eastern station and other stations in the section.

The core of the intermediate maximum salinity in the available stations (1908–1965) from the southern Levant Sea (south of 33° N) and from the northern Levant Sea (north of 35° N) are plotted in two T-S diagrams (Figure 7 and Figure 8). In both regions the winter stations are less frequent.

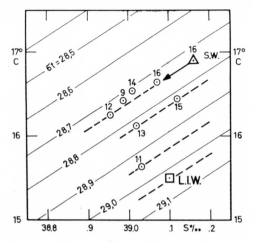

Figure 6 T-S diagram of the core layer of the Levantine intermediate water in
"*Shoyo-Maru*" section, March 1959

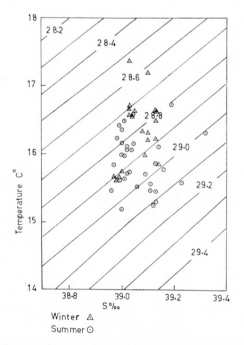

Figure 7 T-S diagram of the core layer of the Levantine intermediate water in the
northern Levant (north of 35° N)

As in Figure 4, both diagrams illustrate that the core of the intermediate layer in summer is cooler than in winter, which suggests that this layer is deeper in summer. By comparing observations made successively in the southern Levant in March (*Shoyo-Maru*), April (*Atlantis*), May (Swedish

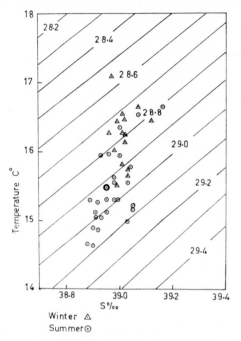

Winter △
Summer ⊙

Figure 8   T-S diagram of the core layer of the Levantine intermediate water in the southern Levant (South of 33° N). "*Ichthyolog*" stations are not included

D.S.E.) and October (*Ichthyolog*), Halim *et al.* (1967) noted a tendency for the intermediate layer to sink gradually from 50–100 m to 200–300 m during the warm season. In addition, the two diagrams (Figures 7 and 8) show that the range of density is similar in both regions, and that the scatter in the points is similarly present. The two diagrams indicate that the conditions in the two regions are not very different.

DISCUSSION AND CONCLUSIONS

There is a general agreement that the northern Levant Sea is the main source of formation of the Levantine Intermediate Water, because: the cold and dry air blowing from Asia Minor, high surface salinity in winter and absence of subsurface minimum salinity during this season. Comparison of climatological factors has shown that the sea surface and overlying air in the northern Levant Sea are only 0.5 °C cooler than in the southern Levant. The waters of the southern Turkish coast are sheltered by the high mountains from the cold dry continental winds. The quantities of rain fall during winter months may reach five- to tenfolds the quantities received by the Egyptian coast.

Air temperature is less than water temperature in both regions during fall and winter indicating favourable conditions for evaporation and vertical mixing. Seasonal variation of surface salinity is mainly affected by evaporation and to a lesser degree by rain fall. Critical examination of available data and published charts for surface salinity indicates opposite trends for the seasonal change of salinity in the northern and southern Levant Sea. In the north the salinity is higher in summer, while it reaches its maximum in winter in the south. This may be attributed, in addition to evaporation, to the heavy rain fall in winter in the north, and to the flow of low salinity water of Atlantic origin which is well developed in summer, and which has a marked influence in this season in the southern Levant compared with its limited effect in the north. Before the erection of Aswan High Dam, the Nile waters have contributed to the decrease of salinity in later summer in the southern region. As a result of these seasonal trends, similar values of surface salinity are observed in both regions in the coldest months of the year, when high salinity contributes to the formation of heavy cold water. Comparison cf spring values shows higher values in the north, since this is the time when the minimum annual salinity occurs in the southern Levant.

A complete absence of subsurface minimum salinity is observed in *"Ichthyolog"* stations off the Egyptian Delta and Sinai in February 1966, and the eastern flank of *"Shoyo-Maru"* section in March 1959. In winter, the Atlantic current becomes feeble, evaporation and winter convection strong enough to eliminate the low salinity east of Alexandria.

Comparison of hydrographic stations in both regions in winter shows similar values of density of the upper layer. It may be slightly lower in the southern region on account of the higher surface temperature. However, *"Shoyo-Maru"* section gives an example of lateral mixing and advective spreading of such water. It is not the absolute high value of density in the whole Levantine basin that initiates sinking and advective processes, but it is the higher values of density relative to the surrounding waters which are similarly slightly warmer than in the north.

"Ichthyolog" observations in 1966 have demonstrated the transformation of water masses in the south east corner of the Levant Sea from the completely homogeneous dense water mass in the upper layer in February to the well developed intermediate (300 m) maximum salinity in August and November. Data accumulating from the southern Levant Sea during the last

decade support the thesis that the southern Levant Sea is a secondary source of formation of intermediate water. Informations from the southern Levant were very scarce, but some evidences supporting the above conclusion can be traced in the literature. The chart representing the distribution of intermediate maximum salinity according to *"Akademik S. Vavilov"* cruise in October 1959 (Moskalenko and Ovchinnikov, 1965, Figure 7), shows three tongues of high salinity of greater than $39.1^{o}/_{oo}$, two larger ones around Rhodes and Cyprus, and a smaller one extending from the Egyptian coast. However, the authors used arrows to show the direction of circulation of the two northern tongues only, apparently neglecting the southern water mass. Similarly the chart of Lacombe and Tchernia (1960, Figure 7) which depends on completely different observations (*Calypso*, October 1956) shows three water masses of greater than $39.0^{o}/_{oo}$; two at the north and a third in the southern Levant Sea. The presence of this isolated water mass at the south cannot be explained by advective processes from the north.

Wüst (1960) defined a water mass having temperature $= 15.5°C$, salinity $= 39.10^{o}/_{oo}$, and $\sigma_t = 29.05$ as a starting point on his T-S diagram (Figure 1) to represent the source region of the intermediate water in winter. Such value was found in the southern Aegean Sea in March 1948, but most available data from the Levant show higher temperatures. In an earlier work, Wüst (1959a) suggested that the points representing source region should be extended vertically down to lower temperatures to reach the higher density of 29.05. Although similar values can be obtained in the severe winter months in the northern and southern Levant Sea, the author is inclined to believe, from evidences at hand, that more than one starting point representing various sources of formation are present in the Levant Sea. These sources differ within a limited range in their salinity, temperature and density. They become closer in their properties as they spread, may be at different levels, towards the Ionean Sea acquiring better homogeneity as they pass the Sicilian ridge to the Tyrrhenian Sea. Wüst's T-S diagram may be modified to show at least two or three sources A_1, A_2 and A_3, which may vary in space and in time, and from which lines extend to meet at a point representing the core layer in the Ionian Sea, which has also a density of 29.05 but a lower temperature and salinity and which lies deeper than in the Levant Sea. This may give a more satisfactory explanation to the large scatter of points representing the intermediate water in the Levant, compared with the remarkable homogeneity of this water in the rest of the Mediterranean.

The data available from the Levantine basin especially in winter are not extensive enough in space or time to show all the features of the distribution of the intermediate water, but it was thought useful to publish the results of this study to show the deficiencies of our present knowledge of this phenomenon. It is hoped that the program of the "Cooperative Investigation of the Mediterranean", which has been recently announced by UNESCO, would take in consideration, when planning future cruises, the need for more closely spaced and more carefully chosen sampling depths to allow the fullest interpretation of the water mass structure of these waters.

Acknowledgements

The author would like to express his gratitude to the National Oceanographic Data Center, Washington, D.C. for providing data used in this paper. He acknowledges the help of his associate H.M.Hassan, M.Sc. (Alexandria) during conducting this work. Sincere thanks are due to Dr. Arnold L. Gordon for his invitation to make this contribution, and for revising the manuscript.

References

Bogdanova, A.K. and M.N.Lebedeva, Some results of the intermediate Levant waters by the observational data obtained in 1960–61 and 1961–62, Institute of Biology of South Seas, A.S. Ukr. S.S.R. Sevastopol, U.S.S.R. (unpublished Manuscript).

Bruce Jr., J.G. and H.Charnock, Studies of winter sinking of cold water in the Aegean Sea, Rapports et Procès-Verbaux des réunions de la C.I.E.S.M.M., **18**, 773, 1965.

El-Kirsh, A.L.M., Study of some hydrographic factors in the waters of Alexandria region, M. Sc. Thesis, 136 pp., Alexandria University, 1969.

El-Maghraby, A.M. and Y.Halim, A quantitative and qualitative study of the plankton of Alexandria waters, *Hydrobiologia*, **25**, Fasc. 1–2, 221, 1965.

Emara, H.I., Distribution of oxygen, nutrient salts and organic matter in the Mediterranean Sea off the Egyptian Coast, M. Sc. Thesis, 190 pp., University of Alexandria, 1969.

Engel, I., Les temperatures dans la Méditerranée orientale, *Cah. Océanogr.*, **18** (6), 507, 1966.

Gorgy, S. and A.H.Shaheen, Survey of U.A.R. Fisheries Grounds, Hydrographic Results of the *Shoyo-Maru* Expedition in the Mediterranean and Red Seas, Oceanographic and Fisheries Research Centre, Alexandria. *Notes and Memoires* No.71, 44 pp., 1964.

Halim, Y., S.K.Guergues, and H.H.Saleh, Hydrographic conditions and plankton in the south east Mediterranean during the last normal Nile flood (1964), *Int. Revue Ges. Hydrobiol.*, **52**, (3), 401, 1967.

Hassan, H.M., The Hydrography of the Mediterranean Waters along the Egyptian Coasts, M. Sc. Thesis, 214 pp., University of Alexandria, 1969.

Lacombe, H. and P.Tchernia, Quelques traits généraux de l'hydrologie Méditerranéenne d'après diverses campagnes hydrologiques rècentes en Méditerranée, dans le proche Atlantique et dans le détroit de Gibraltar, *Cah. Océanogr.*, XII année (8), 527, 1960.

Middelandse Zee, The Mediterranean, Oceanographic and Meteorological data, Royal Netherlands Meteorological Institute, De Bilt, 1957.

Miller, A.R., Physical oceanography of the Mediterranean Sea: A discourse, Rapports et Procès-verbaux des réunions de la C.I.E.S.M.M., **17** (3), 857, 1963.

Mittelmeer-Handbuch, V.Teil, Die Levante, 472 pp., Deutsches Hydrographisches Institut, Hamburg, 1965.

Morcos, S. A., Die Verteilung des Salzgehaltes im Suez-Kanal, Kieler Meeresforsch., **16**, (2), 133, 1960.

Morcos, S. A., On the origin of the Mediterranean intermediate water, IUGG Abstracts of Papers, Vol. V, No. 126, paper presented at International Association of Physical Oceanography, Berne, September 1967.

Morcos, S. A., Physical and Chemical Oceanography of the Red Sea, in *Oceanogr. Mar. Biol. Ann. Rev.*, edited by Harold Barnes, **8** (in press).

Moskalenko, L. V. and I. M. Ovchinnikov, Water masses of the Mediterranean Sea, in "Principal account of the geologic structure, of the régime, and of the biology of the Mediterranean", p. 119, Publication *Nauka*, Moscow, 1965 (in Russian). (Moskva Izdat. Nauka, 1965.)

Nielsen, J. N., Hydrography of the Mediterranean and adjacent waters, *Rap. Dan. Oceanogr. Exped.* 1908–1910, I, 76, Copenhagen, 1912.

Oren, O. H. and I. Engel, Etude hydrologique sommaire du bassin Levantin (Méditerranée Orientale), *Cah. Océanorg.*, **17** (7), 457, 1965.

Ovchinnikov, I. M. and A. F. Fedosejev, On the horizontal circulation of the Mediterranean Sea during summer and winter seasons, in "Principal Account of the geologic structure, of the régime, and of the biology of the Mediterranean", p. 107, Publication *Nauka*, Moscow, 1965 (in Russian). (Académie des Sciences de l'U.R.S.S. 1965.)

Sverdrup, H. U., M. W. Johnson, and R. H. Fleming, The Oceans, their Physics, Chemistry and General Biology, 1087 pp., Prentice-Hall, New York, 1942.

Weather in the Mediterranean, Air Ministry, Vols. I and II, H.M. Stationary Office, London, 1936, 1937.

Weather in the Mediterranean, Meteorological Office, Vol. I, 362 pp., Vol. II, 372 pp., H.M. Stationary Office, London, 1962.

Wüst, G., Remarks on the circulation of the intermediate and deep water masses in the Mediterranean Sea and the methods of their further exploration. Annali Istituto Univ. Navale, **28**, 12 pp., Naples, 1959a.

Wüst, G., Sulle componenti del bilanco idrico fra atmosfera, oceano e Mediterraneo, Annali Istituto Univ. Navale, **28**, 1959b.

Wüst, G., Die Tiefenzirkulation des Mittelländischen Meeres in den Kernschichten des Zwischen- und des Tiefenwassers, Deut. Hydrograph. Z., **13** (3), 105, 1960.

Wüst, G., On the vertical circulation of the Mediterranean Sea, *J. Geophys. Res.*, **66**, 3261, 1961.

Deep Winter-Time Convection in the Western Mediterranean Sea

HENRY STOMMEL*

Muséum National d'Histoire Naturelle
Paris, France

Abstract Selected data obtained in the winter-time Northwestern Mediterranean is presented in such a fashion as to show the time sequence of hydrographic conditions near the center of the region of active deep-water formation. It is shown how quickly deep vertical mixing follows the onset of the Mistral, and how, when it ceases to blow, the mixed water sinks, spreads out horizontally, and unmixed surface strata move in over the region where mixing occurred, sealing it off at the surface. Some comments about the probable physics of the mixing phenomenon are made.

For many years oceanographers have been concerned about the nature of the process by which deep water is formed in the ocean (Wüst, 1928). The regions where deep water is supposed to be formed are infamous for difficult weather and ice conditions, and therefore it has long been recognized that the northwestern Mediterranean—where a local deep-water forms—is a good and accessible region in which to study it (Tchernia, 1956; Tchernia and Saint-Guily, 1959; Wüst, 1960). However the process is complicated in space and time, and it requires several ships working together to survey it adequately.

During the first three months of 1969 the vessels *Hydra Discovery*, *Jean Charcot*, *Origny*, *Atlantis II*, and *Bannock* participated in a multiple ship survey. Because of the mass of material gathered, publication of all the results of these ships' work will not be complete for some years to come. A brief statement of the main results, by the chief scientists of these ships, is already in press (Medoc group, 1970); an account of some of the work

---

* Massachusetts Institute of Technology, Cambridge, Mass. 02139.

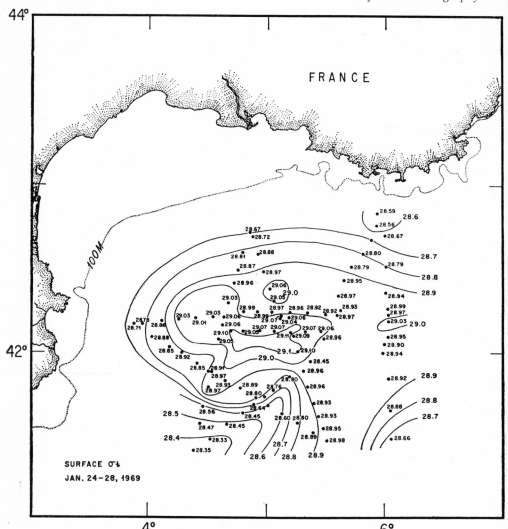

Figure 1 Density (sigma-*t*) at the surface before onset of the Mistral

done by *Atlantis II* (Anati and Stommel, 1970) and of results obtained by floats capable of measuring the vertical component of velocity (Voorhis and Webb, 1970) have already been published.

The present article organizes the data in a somewhat different fashion from that employed in the general article mentioned above (Medoc group, 1970) in an attempt to show in more detail the temporal development of the vertically mixed region, and of the subsequent sinking of the mixture following the cessation of the period of strong cold winds (Mistral).

Even before the onset of the Mistral there is already a region at the surface—about 40 miles south of the French coast—where there is a well-defined density maximum. Figure 1 shows the distribution of surface density (in units of sigma-*t*) as measured by *Atlantis II* during the period 24–28 January 1969. The highest surface densities are centered in a patch at 42° N 5° E, somewhat elongated in the east-west direction. It is in this patch that the deep vertical mixing occurs later during the Mistral.

In order to show the sequence of events in the patch as a function of time and depth, a selection of hydrographic stations and T.S.D. lowerings is made, including only those stations which were made within a small area at the very center of the patch. The choice of stations and their positions are shown in Figure 2.

Figure 3 displays potential temperature, salinity, and density (sigma-theta is a measure of potential density) on this time-depth plane in the neighborhood of 42° N, 5° E. The wet-bulb air temperature (°C) and wind-velocity (knots) measured at the Bouée Laboratoire (42° 15′ N, 5° 30′ E) are also displayed in Figure 3a. There were several periods of strong wind, but the period of greatest wind strength and lowest air temperature was the Mistral of 4–17 February. In the three sections of water properties (Figures 3a, 3b,

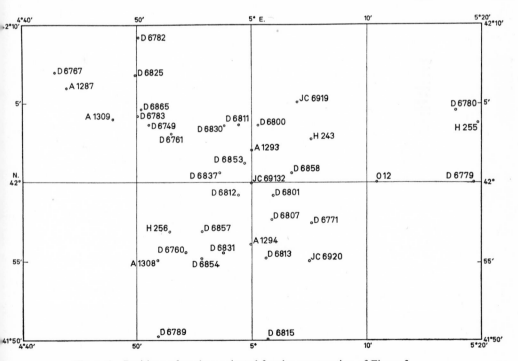

Figure 2   Positions of stations selected for the construction of Figure 3

3c) Nansen casts sample depths are marked with a dot (·). The other stations are T.S.D. lowerings. As can be seen from the section for sigma-theta,
the potential density is steadily stratified all through January. The last stations made in the dense patch before the onset of the Mistral were *Atlantis II*
stations 1308 and 1309 on February 2–3. Due to heavy weather no stations
were made in the area until 7 February when we have *Discovery* station 6749
which shows a mixed layer approximately 1100 meters deep, and an increase
of surface density. It is possible to trace the development of the depth of
the mixed layer in time by noting a rather distinct transition in temperature
and salinity between the values characteristic of the old deep water (potential temperature less than 12.78 °C, and salinity less than 38.420 $^0/_{00}$) and the
values in the warmer, more saline, water of the mixed layer above. Figure 5
shows a salinity vs. potential temperature diagram of *Discovery* station 6761
to indicate how clearly the transition between the two water types is marked.
In Figure 3a the depth of this transition is shown at each station by the position of the dotted line.

At the end of the Mistral the mixed layer has reached its greatest depth
(2150 meters) and the sigma-theta of the mixed layer has become indistin-
guishable from that of the deep water below. The fact that there can be a
transition in temperature and salinity from mixed layer to old deep water,
but no measurable transition in density, can be seen by referring to Figure 5.
We cannot attribute much significance to sigma-theta differences of less than
0.005. Neither of the *Discovery* stations 6779 or 6780 indicate density variations with depth greater than the contour interval 0.005 in sigma-theta.

By 20 February, three days after the Mistral ended, we see that at *Discovery* station 6800 water of lower density has already begun to flow in from
the edges of the narrow convecting region, and now occupies the depth
range 0–800 meters. As time progresses the density stratification rapidly
builds up. By 13 March the stability of the top half of the water column is
essentially restored to its pre-Mistral state. Presumably this rapid restora-
tion of stability is associated with sinking of the mixed water, and spreading
out laterally at depth, and a compensating later inflow of less-mixed water
at the surface. The transition which marks new mixed water's boundary
with old deep water (dotted line in Figure 3a) persists throughout March.
In drawing the contours of Figures 3a, 3b, and 3c properties of very close
pairs of stations have been averaged; these pairs are indicated by brackets
above the station numbers.

Figure 3b exhibits potential temperature on the time-depth plane. During
January there is a marked subsurface temperature maximum (greater than
13.20 °C) associated with the presence of Levantine Intermediate Water.
The surface salinities are relatively low. This distribution persists until the

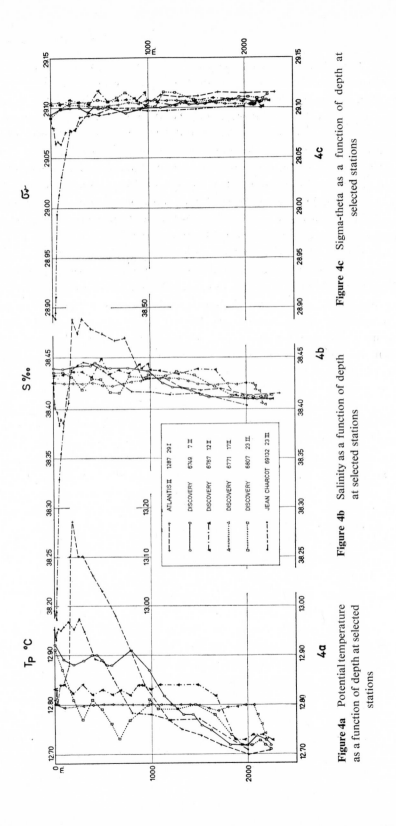

Figure 4a Potential temperature as a function of depth at selected stations

Figure 4b Salinity as a function of depth at selected stations

Figure 4c Sigma-theta as a function of depth at selected stations

T_P °C

S ‰

σ_t

4a

4b

4c

ATLANTIS II	1287	29 I	
DISCOVERY	6749	7 II	
DISCOVERY	6767	12 II	
DISCOVERY	6771	17 II	
DISCOVERY	6807	23 II	
JEAN CHARCOT	69132	23 III	

onset of the Mistral when at *Discovery* station 6749 this maximum disappears because of the strong vertical mixing. The potential temperature of the mixed layer, which at station 6749 is approximately 12.89°C, gradually decreases to 12.81°C at the end of the Mistral. A noticeable increase of surface temperature does not occur until 2 March, and by 8 March a subsurface potential temperature maximum is again in evidence.

The salinity section in Figure 3c also shows the presence of Intermediate Water at the beginning and end of the period—by a somewhat deeper salinity maximum. During the Mistral the upper waters are violently mixed ever deeper with the less saline water below, and the salinity of the mixed layer accordingly decreases from 38.440 to 38.420 °/oo.

Vertical profiles of potential temperature, salinity, and sigma-theta (sigma-*t* computed using potential temperature) are shown in Figures 4a, 4b, and 4c, for each of six stations chosen to illustrate phases in the convection process. Thus *Atlantis II* station 1287 illustrates conditions in the dense patch just before the onset of the Mistral; *Discovery* station 6749 shows mixing to about 900 meters after 3–4 days of intense Mistral; station 6767 when mixing extends to about 1750 meters; and at the very end of the Mistral, *Discovery* station 6771 shows remarkable homogenization down to 2150 meters. Within six days of the end of the storms large mass movements have evidently occurred in the area, there is no single mixed surface layer—as shown by *Discovery* station 6807. A month later, *Jean Charcot* station 69132 shows that both at the surface and in the deep water the structure is returning to the initial conditions. Between 800 meters and 1500 meters there appears to remain a thick layer of the newly mixed water.

Examination of the temperature, salinity and sigma-theta scales will indicate to the reader the unusually small range of variation of properties. Thus, for example, the range of salinity in most stations displayed in Figure 4b is within the limits of error of the old chemical titration method of salinity determination (our determinations were made with modern salinometry), and the differences in computed sigma-theta (Figure 4c) of the deep water appear to be a function of the observing vessel, rather than of anything else.

Comparison of *Atlantis* station 1287 and *Discovery* station 6771 gives some idea of the total heat loss to the water column from just before the onset of the Mistral (29 January) to the end (17 February). Reference to Figure 4a shows that there was cooling between the surface and 1050 meters, warming between 1050 meters and 2200 meters. Numerical integration of temperature over the total depth indicates a net heat loss of 8200 gm calories cm^{-2} during the duration of the Mistral. There is some evidence that *Discovery* temperatures should be reduced by 0.01°C (or *Atlantis* and *Charcot* temperatures increased by 0.01 °C) to make deep temperatures taken by all vessels comparable. If this is done the calculated loss of heat is computed to be 10,400 gm cals cm^{-2}. A glance at Figure 4b indicates that sampling errors will probably vitiate any attempts to compute net salinity change. Carrying out a numerical integration of the salinity for each of the same two stations indicates a net decrease of salt content of approximately 2×10^{4} $^{0}/_{00}$ cm, whereas due to a net evaporation of the order of 10 cm we might have expected an increase of something like 4×10^{2} $^{0}/_{00}$ cm. It is evident that this figure is far too small to be determined by our data, a point made also by Anati and Stommel in their discussion of *Atlantis* data (1970).

Changes of the density with time are important to a dynamical understanding of the convection process (Kraus and Turner, 1967). The scatter in Figure 4c indicates that vertical integrals are likely to be quite uncertain, even attempts to determine density changes at fixed depth with time are bound to be uncertain. In order to determine whether the theory of Kraus and Turner is possibly applicable it is necessary that as the depth of the

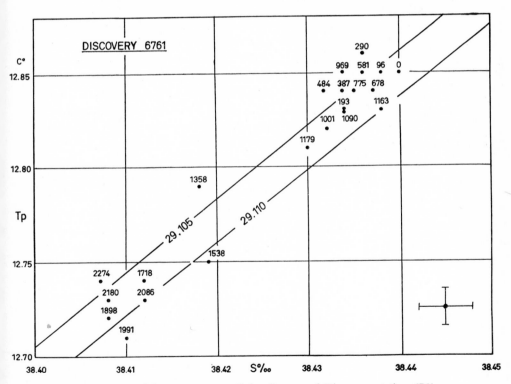

Figure 5 A potential temperature—salinity diagram of Discovery station 6761 showing the sharp transition between two water types characteristic of the boundary between the mixed layer, and old deep water

mixed layer passes any depth, the density at that depth should initially decrease (evidence of penetrative convection). The truth or falsity of this critical fact can scarcely be determined by comparing individual density profiles as shown in Figure 3c. Accordingly, a statistical approach is indicated, and we have divided stations where the convective front is clearly marked by the change in water mass properties into four groups: Group A, with observed mixed layer depth less than 1200 meters, Group B with mixed layer depth between 1200 and 1600 meters, Group C with mixed layer depth between 1600 meters and 2000 meters, and finally Group D with mixed layer depth greater than 2000 meters. For each of these groups we have computed the mean density (sigma-theta) at 1200, 1600, and 2000 meters. The results are shown in Table 1. The sigma-thetas are computed to 1 part in ten thousand (but this is certainly well beyond their significance because the standard deviation of individual values is something like two parts per thousand, and since no means are computed from more than 13 individual values, they all have fiducial limits of approximately 0.0007). At 1200 meters, sigma-theta in Group A is 29.1020, and in Group B is 29.1048, which appears to be a significant increase instead of a decrease as might have been expected in a process of penetrative convection. Similar increases occur at 1600 meters and 2000 meters.

Our next question is to ask why the region of strong convection is so small geographically. In order to explore this we must choose data which is well distributed in depth, time, and latitude. This is somewhat different

Table 1

Number of stations	Group A	Group B	Group C	Group D	
	7	13	13	9	
Mixed layer depth (meters)	700–1150	1220–1560	1700–1950	2030–2240	
Depth					
1200	29.1020	29.1048	29.1068	29.1093	
1600	29.1050	29.1050	29.1075	29.1090	mixed
2000	29.1087	29.1076	29.1078	29.1090	
					boundary

unmixed

0.0020 individual deviation

Station list: Group A 6746, 49, 53, 54, 57, 60, 94
              Group B 6752, 58, 61, 62, 63, 64, 68, 69, 74, 75, 81, 93, 6817
              Group C 6759, 66, 67, 72, 76, 77, 78, 84, 88, 6807, 10, 12, 16
              Group D 6779, 80, 83, 85, 86, 90, 6800, 01, 13

from the choice made earlier, where the distribution was in depth and time, with both latitude and longitude fixed to a small area at the center of the dense patch. The only place where we have this distribution of data is at 6° E where a meridional section was occupied five times by *Atlantis II* and *Jean Charcot*. Unfortunately this longitude is somewhat to the east of the center of the dense patch, but it seems as though it will do satisfactorily for the present purpose. The sections are long enough in latitude to embrace not only the regions where strong convection occurs, but also regions to the north and south where vertical convection is severely limited in depth.

Table 2 Stations used in Figure 4, on 6° E longitude

Latitude N	*Atlantis II*	*Jean Charcot*	*Atlantis II*
42°40′	1285 26 I	6945 17 II	1399 4 III
42°30′	1284 25 I	6946 17 II	1400 4 III
42°20′	1283 25 I	6947 17 II	1401 4 III
42°10′	1282 25 I	6949 18 II	1403 5 III
42°00′	1281 25 I	6951 18 II	1405 5 III
41°50′	1280 25 I	6966 20 II	1406 5 III
41°40′	1279 24 I	6958 18 II	1408 5 III
41°30′	1278 24 I	6959 19 II	1409 5 III
41°20′	1277 24 I	6960 19 II	1410 5 III

The stations chosen are listed in Table 2 according to latitude at three different periods: *Atlantis II* stations 1277–1285 before the Mistral (23–26 January); *Jean Charcot* stations 6945–51, and 6956, 58, 59, 60, 66, immediately after the storm (17–19 February); and *Atlantis II* stations 1399–1410 two weeks later (4–6 March). For each of these stations the difference in average sigma-theta in the upper 1000 meters and that between 1000 meters and 2000 meters was computed according to the formula

$$\Delta\sigma = \left\{ \int_{1000}^{2000} \sigma_\theta dZ - \int_0^{1000} \sigma_\theta dZ \right\} \Big/ 1000$$

This quantity is displayed in the $\Delta\sigma$ vs latitude diagram shown in Figure 6.

From the convention of signs employed, positive values of $\Delta\sigma$ are a measure of the buoyancy in the upper 1000 meters. Curve (1) in Figure 6 thus represents the mean buoyancy in the upper layers as a function of latitude before the onset of the Mistral; whereas curve (2) represents the mean buoyancy immediately following the end of the Mistral. Two features are clearly evident. First, the initial buoyancy is not uniform, but is least in the middle: between latitudes 41° 50′ N and 42° 20′ N. During the Mistral buoyancy is lost *more or less uniformly* at all latitudes (except at the northernmost stations where the loss is smaller). Offhand this seems somewhat surprising; one

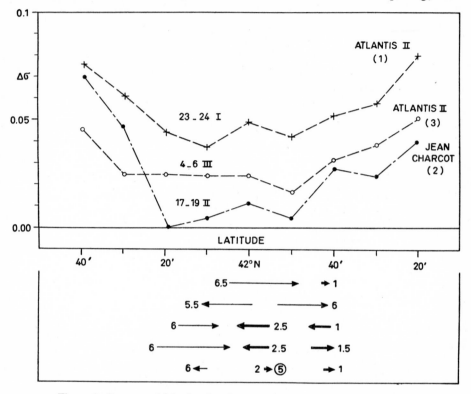

Figure 6   Buoyancy ($\Delta\sigma$) of surface layer at 6° E as a function of latitude at three times. Also north–south displacements of floats (see text)

might expect greatest cooling and evaporation nearest to the land where the cold wind originates, unless fetch is more important than nearness to the cold source. The average buoyancy loss (change in $\Delta\sigma$) during the period of the Mistral (4–17 February) is 0.035. Rough calculation of expected evaporation and heat loss (from the meteorological data of the Bouée Laboratoire) indicates an evaporation of approximately 3 cm day^{-1} and an associated heat loss of at least 2000 gm cal cm^{-2} day^{-1}, and hence a daily decrease in $\Delta\sigma$ of 0.005. The observed uniform decrease is therefore consistent with 7 days of strong Mistral. Inasmuch as there were 5 or 6 days of low winds during the period 4–17 February, the computed buoyancy loss is consistent with that observed.

The second feature of importance to notice is that only in the central latitudes is the buoyancy reduced essentially to zero, so that this is the only place where very deep convective mixing can occur. The implication is that the smallness of the convective region is not due to a localization of meteorological nature, nor to hydrodynamical mechanisms such as those studied by Stommel (1962) and Rossby (1965), but depends solely upon the distribution of buoyancy prior to the Mistral. Since this latter seems to depend upon the general circulation of this region of the Mediterranean which in turn is not quantitatively understood (a qualitative theory has been advanced by Saint-Guily (1963)), we cannot pursue the explanation further. Many of the features of deep-water formation observed in the Mediterranean are similar to those surmised to occur in the Norwegian Sea (Mosby, 1967).

Some of the above discussion is based upon the assumption that during the violent mixing of the Mistral there are no large movements of water in the north-south direction. At 6° E we do not have adequate direct measurements of current from which we can determine the actual horizontal displacements which occurred. There are, however, some float tracks made at depths near 500 meters by *Discovery* near 5° E and these give at least an indication of the magnitude of the displacements to be expected at 6° E. The arrows in the lower half of Figure 6 indicate the beginning and ending latitudes of the drift of these floats. The numerals indicate the number of days elapsed during the displacement as measured; heavy arrows indicate floats tracked during 9–14 March; the light arrows, during 23–28 February. The spot with the numeral 5 indicates a float which had zero displacement in latitude after 5 days' drift. As can be seen the real displacements of water masses by these float measurements are substantial. We do not have tracks during the period of intense mixing, but if the displacements were much larger it would be impossible to draw some of the quantitative inferences which we have. Of course we do attribute the increase of buoyancy as indicated by the general rise in $\Delta\sigma$ from curve (2) to curve (3) to horizontal movements of water from regions outside of the narrow convective zone which presumably accompany the sinking and spreading of the mixed water after the end of the Mistral.

The author has been able to write this paper only because of the efforts at sea of his many colleagues, in particular John Swallow, Paul Tchernia, Henry Lacombe, Arthur Miller, and Roberto Frassetto. He is also indebted for support to the John Simon Guggenheim Memorial Foundation, contracts GA-1613 and GA-12773 from the U.S. National Science Foundation, and to the Laboratoire d'Océanographie Physique du Muséum National d'Histoire Naturelle, Paris.

References

Anati, David and Henry Stommel, The initial phase of deep-water formation in the North-west Mediterranean, during Medoc '69, on the basis of observations made by *Atlantis II*, January 25, 1969 to February 12, 1969. *Cahiers Océanographiques*, XXII, 4, pp. 343–351, 1970.

Kraus, E. B. and J. S. Turner, A one-dimensional model of the seasonal thermocline. II. The general theory and its consequences. *Tellus* XIX, pp. 98–105, 1967.

Medoc group, Observation of formation of deep water in the Mediterranean Sea, 1969. Nature, 227, pp. 1037–1040, 1970.

Mosby, H., Fridtjof Nansen, Minnesforelesningen III. *Det Norske Videnskaps-Akademi i Oslo*, 29 pp., 1967.

Rossby, H. T., On thermal circulation driven by non-uniform heating from below. *Deep-Sea Res.* (12) pp. 9–16, 1965.

Saint-Guily, B., Remarques sur le Mécanism de Formation des eaux profondes en Méditer-ranée Occidentale. *Rapp. et Procés-verbeaux CIESMM* (XVII) 3, 1963.

Stommel, Henry, On the smallness of sinking regions in the ocean. *Proc. Nat. Acad. Sci.* Wash. 48, pp. 766–772, 1962.

Tchernia, P., Contribution à l'étude hydrologique de la Méditerranée occidentale. *Bull. d'Information du Comité Central d'Océanographie et d'Etude des Côtes*, VIII, 9, pp. 427–463, 1956.

Tchernia, P. and B. Saint-Guily, Nouvelles observations hydrologiques d'hiver en Méditer-ranée Occidentale. *Cahiers Oceanographiques du COEC*. XI. 7, pp. 499–542, 1959.

Voorhis, A. and D. C. Webb, Large Vertical Currents observed in a Western Sinking Region of the Northwestern Mediterranean. *Cahiers Océanographiques*, xxii, 6, pp. 571–580, 1970.

Wüst, G., Die Tiefenzirkulation des Mittelländischen Meeres in den Kernschichten des Zwischen- und des Tiefenwassers. *Deutsche Hydrogr. Zeitschr.* **13**, 3, pp. 105–131, 1960.

Wüst, G., Der Ursprung der Atlantischen Tiefenwasser, Jubiläums-Sonderband, *Zeitschr. Ges. f. Erdkunde* Berlin, pp. 507–534, 1928.

Reproduction of Currents and Water Exchange in the Strait of Gibraltar with Hydrodynamical Numerical (HN) Model of Walter Hansen

TAIVO LAEVASTU

Environmental Prediction Research Facility
Monterey, California

Abstract   The Hydrodynamical Numerical (HN) model of Walter Hansen has been used to compute tides and currents in the Strait of Gibraltar. The model has two open boundaries at which the tides were prescribed at each time step, using four tidal constituents. The grid size was two nautical miles, and the time step (from Courant criterion) was 24 seconds. Equilibrium was established after 20 hours of real-time computation. The main results from these model runs were: (1) At mean tides, the inflowing currents are stronger and last somewhat longer than the outflowing currents. (2) There is a net inflow into the Mediterranean Sea caused by tidal currents alone (excluding wind and sea level difference). (3) The tides computed with the HN model were in good agreement with those obtained from the harmonic method at points where harmonic constants were available. There was also a good agreement between computed currents and the few available current measurements. (4) The HN model gave a net transport of 1.48 km^3 per hour into the Mediterranean in the upper 100 m. This value is somewhat higher than previous estimates; however, heat budget considerations require a higher water exchange through the Strait of Gibraltar than given by previous estimates.

INTRODUCTION

The Hydrodynamical Numerical (HN) models of Professor Walter Hansen, University of Hamburg, have been tested on a number of shallow semi-closed seas. The model closely reproduces observed tides, storm surges and currents (Hansen, 1966; Sündermann, 1966; Laevastu and Stevens, 1969). It was desirable to test this model on the Strait of Gibraltar, which is a relatively deep area and has two open boundaries with distinctly different tides.

It is of interest to know whether the latter aspect is properly reproduced by HN models. Furthermore, the Strait of Gibraltar is an ideal area for eventual testing of a two-layer HN model. This report presents the results and some verifications of single-layer HN model tests for the Strait of Gibraltar.

HYDRODYNAMICAL FORMULAS AND W. HANSEN'S FINITE DIFFERENCE METHOD

The hydrodynamical numerical method for computation of currents and sea level changes was proposed in its present form by Hansen in 1956.

The following basic equations are used in the single-layer model:

$$\frac{\partial u}{\partial t} - fv - v\,\varDelta u + \frac{r}{H}\,u\sqrt{u^2 + v^2} + g\,\frac{\partial \zeta}{\partial x} = X + \frac{\tau^{(x)}}{H} \tag{1}$$

$$\frac{\partial v}{\partial t} + fu - v\,\varDelta v + \frac{r}{H}\,v\sqrt{u^2 + v^2} + g\,\frac{\partial \zeta}{\partial y} = Y + \frac{\tau^{(y)}}{H} \tag{2}$$

$$\frac{\partial \zeta}{\partial t} + \frac{\partial}{\partial x}\,(Hu) + \frac{\partial}{\partial y}\,(Hv) = 0 \tag{3}$$

$\tau^{(x)}$ (and $\tau^{(y)}$) are usually expressed as:

$$\tau^{(x)} = \lambda W_x \sqrt{W_x^2 + W_y^2} \tag{4}$$

The bottom stress (friction) term in formulas 1 and 2 is:

$$\tau^{(b)} = \frac{r}{H}\,u\sqrt{u^2 + v^2};\quad \frac{r}{H}\,v\sqrt{u^2 + v^2} \tag{5}$$

The following symbols were used in the formulas above:

x, y	space coordinates
t	time
u, v	components of water velocity
H	total depth ($H = h + \zeta$)
ζ	surface elevation above a level sea surface
X, Y	components of external forces
$\tau^{(x)}, \tau^{(y)}$	components of wind stress
g	acceleration of gravity

f Coriolis parameter

r friction coefficient (3×10^{-3}) for bottom stress

ν coefficient of horizontal eddy viscosity

∇ Laplace operator

λ coefficient of friction or surface drag (3.5×10^{-6})

W_x, W_y wind speed components

$\tau^{(b)}$ bottom stress

Analytical solution(s) to formulas 1 to 3 are of little value, as exact solutions are possible only for basins of regular shape, simple depth and simple wind distribution. However, the formulas can be solved on computers using "step-by-step" finite-difference methods.

$$\zeta^{t+\tau}(n,m) = \xi^{t-\tau}(n,m)$$

$$- \frac{\tau}{l} \{H_u^t(n,m)\,U^t(n,m) - H_u^t(n,m-1)\,U^t(n,m-1)$$

$$+ H_v^t(n-1,m)\,V^t(n-1,m) - H_v^t(n,m)\,V^t(n,m)\} \qquad (6)$$

$$U^{t+2\tau}(n,m) = \{1 - [2\tau r/H_u^{t+2\tau}(n,m)]\sqrt{\bar{U}^t(n,m)^2 + V^{*t}(n,m)}\}\,\bar{U}^t(n,m)$$

$$+ 2\tau f V^{*t}(n,m) - \frac{\tau g}{l}\{\zeta^{t+\tau}(n,m+1) - \zeta^{t+\tau}(n,m)\}$$

$$+ 2\tau X^{t+2\tau}(n,m) \qquad (7)$$

$$V^{t+2\tau}(n,m) = \{1 - [2\tau r/H_v^{t+2\tau}(n,m)]\sqrt{\bar{V}^t(n,m)^2 + U^{*t}(n,m)^2}\}\,\bar{V}^t(n,m)$$

$$- 2\tau f U^{*t}(n,m) - \frac{\tau g}{l}\{\zeta^{t+\tau}(n,m) - \zeta^{t+\tau}(n+1,m)\}$$

$$+ 2\tau Y^{t+2\tau}(n,m) \qquad (8)$$

The "averaged" velocity and water elevation (sea level) components are:

$$\bar{U}^t(n,m) = \alpha U^t(n,m) + \frac{1-\alpha}{4}\{U^t(n-1,m) + U^t(n+1,m)$$

$$+ U^t(n,m+1) + U^t(n,m-1)\} \qquad (9)$$

$V^t(n,m)$ and $\bar{\zeta}^t(n,m)$ are analogous.

(The factor α can be interpreted as "horizontal visocsity parameter". Its normal value is 0.99 (see Laevastu and Stevens, 1969).)

$$U^{*t}(n,m) = \tfrac{1}{4}\{U^t(n, m-1) + U^t(n+1, m-1)$$

$$+ U^t(n,m) + U^t(n+1, m)\} \tag{10}$$

$V^{*t}(n,m)$ is analogous to the $U^{*t}(n,m)$ above.

The time step is 2τ. The total depth (H_u, H_v) is computed as

$$H_u^{t+2\tau}(n,m) = h_u(n,m) + \tfrac{1}{2}\{\zeta^{t+\tau}(n,m) + \zeta^{t+\tau}(n, m+1)\} \tag{11}$$

The effects of wind (external force) are computed with the following formula for X component

$$X^t = \frac{\lambda W_x^t \sqrt{(W_x^t)^2 + (W_y^t)^2}}{H} - \frac{1}{\varrho r}\frac{\partial p_0}{\partial x} \tag{12}$$

and for the Y components:

$$Y^t = \frac{\lambda W_y^t \sqrt{(W_x^t)^2 + (W_y^t)^2}}{H} - \frac{1}{\varrho r}\frac{\partial p_0}{\partial y} \tag{13}$$

A number of slightly modified finite difference schemes are possible for solving the hydrodynamical equations. Some of these schemes are being tested at Fleet Numerical Weather Center (FNWC).

HYDRODYNAMICAL NUMERICAL (HN) MODEL OF STRAIT OF GIBRALTAR

The computational grid for the Strait of Gibraltar has a two nautical mile mesh length. This grid size and the maximum depth in the area require (according to Courant, Friedrich, Lewy criterion) a time step of 24 seconds in order that the computations remain stable. Depths at u and v grid points were obtained from navigation charts. The friction coefficient was 0.003. Tides with four tidal constituents $(M_2, S_2, N_2$ and $K_2)$ were introduced at each time step at both ends. The tidal input was constant across the openings. It has been found at FNWC that the tidal input at the openings can be made an inverse function of depth. However, if this is not done, the program will adjust the z and u, v values a few gridpoints inwards. Equilibrium in this particular program was established in about 20 hours of real time. The program was run 100 hours. Mean tides (tides between spring and neap) and the hourly outputs were taken from the last 25 hours of computation. The program was also run with wind inputs, but essentially calm wind results are reported in this paper.

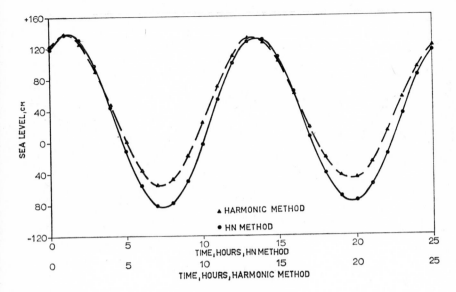

Figure 1 Tides of Tangier, computed with HN and harmonic methods
(M_2, S_2, N_2, K_2)

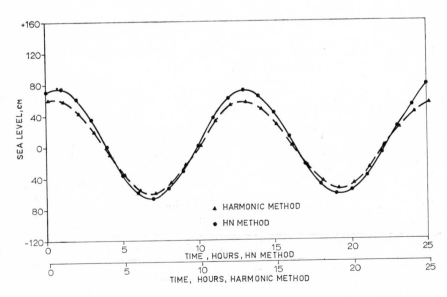

Figure 2 Tides at Tarifa, computed with HN and harmonic methods
(M_2, S_2, N_2, K_2)

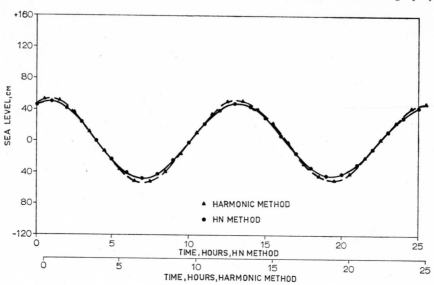

Figure 3   Tides at Gibraltar, computed with HN and harmonic methods
(M_2, S_2, N_2, K_2)

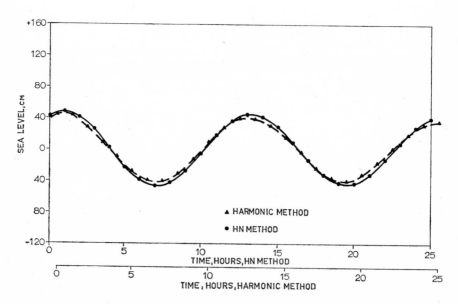

Figure 4   Tides at Ceuta, computed with HN and harmonic methods
(M_2, S_2, N_2, K_2)

There are some limitations in the application of HN model to the Strait of Gibraltar area, arising from the formulation of the original model. First the area is a pronounced two-layer system; secondly it is a deep area; and thirdly it has two distinctly different open boundaries. In the application of the model in the Strait of Gibraltar area some steps were taken to alleviate these limitations and the results should reflect partly the success of these specific steps: The tidal currents were computed from surface to bottom, but wind effects only to a depth of 100 meters. The second boundary conditions (on the Mediterranean side) were presented by using available tidal constituents in the vicinity of this second boundary, but otherwise the second boundary was left "open" for u and v current components.

VERIFICATION OF SEA LEVEL COMPUTATIONS

The tides at a number of locations for which tidal harmonic constants were available were computed with the well-known harmonic method. The same time span as the output from the HN model was used as well as the same four tidal constituents. The sea level changes computed with the HN method were taken from a grid point closest to the original location of the corresponding tide gauge constituents used in the harmonic computations. Some comparisons are shown in Figures 1 to 4.

At Tangier (Figure 1), there is virtually no time lag between the HN and harmonic models. However, the HN model gives about 10% higher amplitude, the difference occurring mainly in low water prediction. At Tarifa (Figure 2), the tides with the HN method precede about 10 minutes the tides from the harmonic method and the HN tides are about 8% higher. At Gibraltar (Figure 3), and at Ceuta (Figure 4), the tidal heights of both methods are in relatively good agreement; however, the HN tides precede the "harmonic" tides by about 30 minutes. It should be noted that the HN tides are taken from grid points which are as much as two miles away from the tide gauges and usually over deeper water.

VERIFICATION OF COMPUTED CURRENTS

In addition to producing tidal forecasts in areas where harmonic constants are not available, the HN model has the advantages of handling superimposed meteorological tides (storm surges) and of predicting tidal currents. The latter were not obtainable using "harmonic" methods unless current recordings were available over relatively long intervals.

Examples of synoptic tidal currents in the Strait of Gibraltar are given in Figures 5 and 6. The direct verification of synoptic currents is difficult due

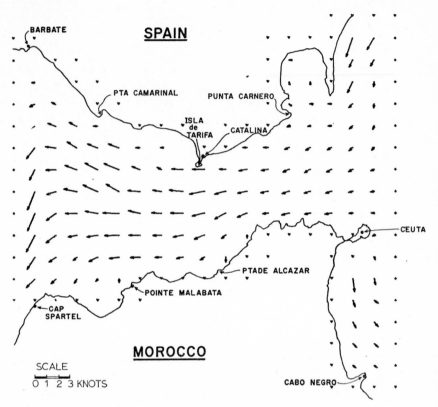

Figure 5 Tidal currents in Strait of Gibraltar three hours after low water at Tarifa. (HN method)

to the scarcity of current measurements. Some quasi-synoptic current charts have been prepared subjectively using a variety of data, including ships' logbooks. Gaps have often been filled in using some degree of "artistic license". Two subjective tidal current charts for the Strait of Gibraltar are given in Figures 7 and 8 (from NOO Publ. 700, 1965), corresponding to the times and currents in Figures 5 and 6. Although detailed comparison of the two sets of charts has no direct use, it can be noted that agreement is good.

A more detailed comparison of measured and computed tidal currents at a given location in the western entrance to the Strait of Gibraltar is given in Figure 9. Although the exact stage of tides (mean, spring or neap) during the measurement is not known, the relation between the observed and computed tides is a good one. The tidal currents, computed with the HN method are about 20% stronger.

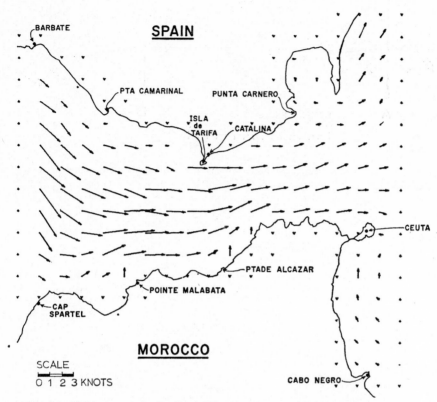

Figure 6 Tidal currents in Strait of Gibraltar three hours after high water at Tarifa. (HN method)

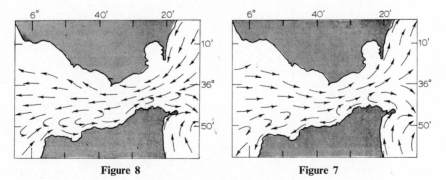

Figure 7 Tidal currents in Strait of Gibraltar three hours before high water at Gibraltar (subjective summary, NOO Publ. 700)

Figure 8 Tidal currents in Strait of Gibraltar three hours after high water at Gibraltar (subjective summary, NOO Publ. 700)

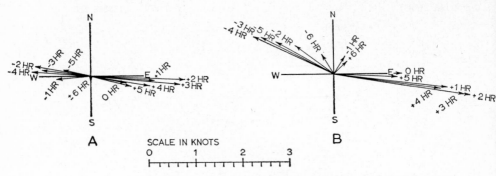

Figure 9a Measured tidal currents at 250 m at 35° 54′ N, 5° 52′ W (from NOO Publ. 700.) B. Currents computed with HN method at about the same location. Times indicate hours before and after high water at Gibraltar

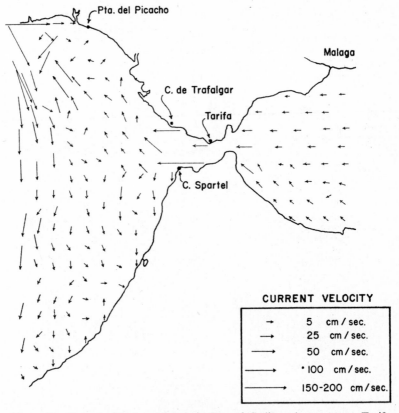

Figure 10 Currents (tidal and wind) in Bay of Cadiz at low water at Tarifa, computed with HN model (wind 6 m sec^{-1} from E)

It should be noticed that the tidal currents are considerably weaker in the Bay of Cadiz than at the entrance to and inside the Strait of Gibraltar. An example of currents in the Bay of Cadiz computed with another HN model is given in Figure 10.

TRANSPORT OF WATER THROUGH THE STRAIT OF GIBRALTAR

The general circulation in the Strait of Gibraltar (inflow of Atlantic water at the surface, outflow of Mediterranean water along the bottom) has been discussed in a number of publications and various quantitative estimates of the water exchange have been given (e.g. Wüst, 1952, 1959; Lacombe, 1961,

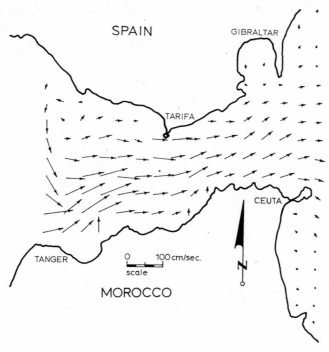

Figure 11 Tidal rest currents in Strait of Gibraltar (HN model)

and others). The possible influence of different forces in effecting the seasonal changes of the water exchange, such as atmospheric pressure difference, winds, difference in evaporation in the Mediterranean, etc., have also been discussed. However, as far as can be ascertained by the author, the net (or rest) tidal circulation in the water exchange in the Strait of Gibraltar has not been pointed out earlier.

Due to the variation of amplitude and phase of the tide over some distance, there will be a net transport of tidal currents. This net circulation over a mean tidal cycle (24 h 50 min) in the Strait of Gibraltar is shown in Figure 11. As seen from this figure, the tides alone might effect the net inflow to the Mediterranean. This is due to the amplitude difference at opposite ends of the straits.

The best measurements of currents in the Strait of Gibraltar have been made by Lacombe (1961). The net currents, computed from these measurements, are shown in Figures 12 and 13. As seen, the main inflow occurs above 100 m depth.

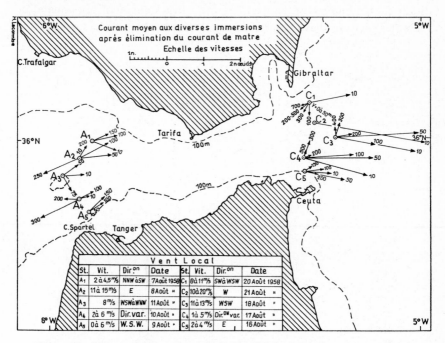

Figure 12　Measured rest currents in Strait of Gibraltar (after Lacombe 1961).

The transport of water through the straits from 0 to 100 m during a tidal cycle was extracted from the HN model and is presented in Figure 14. The net tidal inflow is 1.48 km^3/hour (0.4×10^6 m^3/sec). Note that this value for the inflow does not include the thermohaline component to the velocity field. This value is about half of that obtained by Lacombe from direct measurements (2.7 km^3 per hour), but about ten times as high as Lacombe's adjusted value (0.15 km^3 per hour).

If the net evaporation per square centimeter in the Mediterranean is about 150 cm/year, and all the water deficit were to be balanced by the

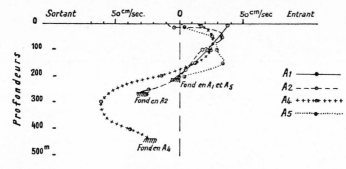

Composantes de courant moyen normales aux sections

Section de l'Ouest A₁-A₅ (sauf A₃)

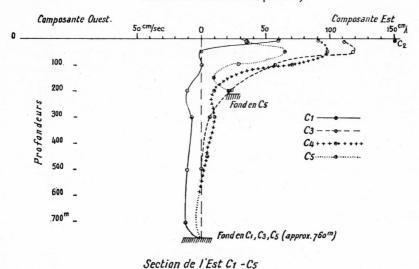

Section de l'Est C₁-C₅

Figure 13 Measured rest currents at different depth in Strait of Gibraltar (locations see fig. 12. after Lacombe 1961).

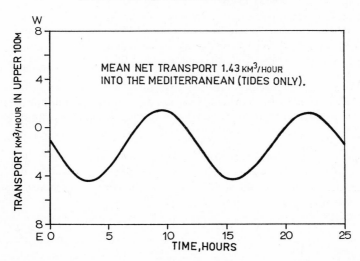

Figure 14 Transport of water through Strait of Gibraltar from surface of 100 m (HN model)

inflow through the Strait of Gibraltar, a maximum net inflow of 0.45 km³ per hour would be required. Furthermore, heat budget computations at FNWC for the Mediterranean (Laevastu *et al.*, 1970) require a larger water exchange than heretofore estimated. More accurate insight into these problems is expected from the two-layer HN model, which would include aspects of the thermohaline flow.

SUMMARY

a   The HN model with two open boundaries can be used for computation of tides and currents (including wind currents) in relatively deep straits.

b   The tidal phases and amplitudes computed with the HN model for the Strait of Gibraltar verify well. The maximum difference has been 30 minutes in time of high water and 20% in amplitude.

c   The net (or rest) tidal current gives a net inflow into the Mediterranean. The inflowing tidal currents are stronger and also last somewhat longer than the outflowing tidal currents.

d   The computed net inflow is somewhat larger than previously available estimates. However, heat budget computations require a large water exchange. Multilayer HN models will shed further light on this problem.

References

Hansen, W., Theorie zur Errechnung des Wasserstandes und der Strömungen in Randmeeren nebst Anwendungen. *Tellus*, **8**, 287–300, 1956.

Hansen, W., The reproduction of the motion in the sea by means of hydrodynamical-numerical methods. *Mitteil. Inst. Meeresk., Hamburg*, **5**, 57 pp. plus figures, 1966.

Lacombe, H., Contribution à l'étude du régime dué dtroit de Gibraltar. I—Etude dynamique. *Cah. Océanogr.*, **13** (2), 73–107, 1961.

Laevastu, T. and P. Stevens, Application of hydrodynamical-numerical models in ocean analysis/forecasting. Part I—The single-layer models of Walter Hansen. *FNWC Techn. Note No. 51*, 45 pp. plus figures, 1969.

Laevastu, T., L. Clark, and P. Stevens, Annual cycles of heat in the northern hemisphere oceans and heat distribution by ocean currents. FNWC Tech. Note No. 53, 1970.

Sündermann, J., Ein Vergleich zwischen der analytischen und der numerischen Berechnung winderzeugter Strömungen und Wasserstände in einem Modellmeer mit Anwendung auf die Nordsee. *Mitteil. Inst. Meeresk., Hamburg*, **4**, 73 pp. plus tables, 1966.

U.S. Naval Oceanographic Office, Oceanographic atlas of the North Atlantic Ocean. Section I—Tides and currents. *NOO Publ. No. 700*, 75 pp., 1965.

Wüst, G., Der Wasserhaushalt des Mittelländischen Meeres und der Ostsee in vergleichender Betrachtung. *Geof. pura e. appl.*, **21**, 7–18, 1952.

Wüst, G., Sulle componenti del bilancio idrico fra atmosfera oceano e Mediterraneo. *Ann. Inst. Univ. Navale, Napoli*, **28**, 1–18, 1959.